TRADITION, CULTURE AND
IN AFRICA

Heritage, Culture and Identity

Series Editor: Brian Graham,
School of Environmental Sciences, University of Ulster, UK

Other titles in this series

Ireland's Heritages
Critical Perspectives on Memory and Identity
Edited by Mark McCarthy
ISBN 0 7546 4012 4

Senses of Place: Senses of Time
Edited by G. J. Ashworth and Brian Graham
ISBN 0 7546 4189 9

(Dis)Placing Empire
Renegotiating British Colonial Geographies
Edited by Lindsay J. Proudfoot and Michael M. Roche
ISBN 0 7546 4213 5

Preservation, Tourism and Nationalism
The Jewel of the German Past
Joshua Hagen
ISBN 0 7546 4324 7

Tradition, Culture and Development in Africa
Historical Lessons for Modern Development Planning

AMBE J. NJOH
University of South Florida, USA

Routledge
Taylor & Francis Group

LONDON AND NEW YORK

First published 2006 by Ashgate Publishing

2 Park Square, Milton Park, Abingdon, Oxon OX14 4RN
711 Third Avenue, New York, NY 10017, USA

Routledge is an imprint of the Taylor & Francis Group, an informa business

First issued in paperback 2016

British Library Cataloguing in Publication Data
Njoh, Ambe J.
    Tradition, culture and development in Africa : historical
    lessons for modern development planning. - (Heritage,
    culture and identity)
    1. Economic development - Social aspects - Africa,
    Sub-Saharan 2. Africa, Sub-Saharan - Social conditions
    I. Title
    306'.0967

Library of Congress Cataloging-in-Publication Data
Njoh, Ambe J.
    Tradition, culture and development in Africa : historical lessons for modern
development planning / by Ambe J. Njoh.
       p. cm. (Heritage, culture and identity)
    Includes index.
    ISBN-13: 978-0-7546-4884-0
    ISBN-10: 0-7546-4884-2
  1. Africa--Economic conditions. 2. Africa--History. 3. Africa--Civilization. I. Title. II. Series

    HC800.Z9E4465 2006
    338.96--dc22

              2006023908

ISBN 13: 978-0-7546-4884-0 (hbk)
ISBN 13: 978-1-138-26269-0 (pbk)

Contents

To the memory of Mama Elizabeth Njweng Njoh, the only 'father' I ever knew.

List of Figures and Tables

Preface and Acknowledgements

The need to debunk theories that incriminate indigenous customs, culture and traditional practices as causes of Africa's underdevelopment is long overdue. I first entertained the idea of interrogating such theories when I became acquainted with the recommendations of development economists of the late-1950s and early-1960s, such as Walt Rostow, who considered the cultural transformation of Africa and other developing regions a *conditio sine qua non* for economic development. For these economists and cognate thinkers, the customs and traditional practices of non-Western societies, constitute a hurdle to so-called modern development aspirations.

Classical economists and proponents of modernization view problems such as administrative ineptitude, bureaucratic corruption, economic underdevelopment, and technological backwardness that constitute defining characteristics of contemporary African countries, as products of African indigenous customs, culture and tradition. Outlandish charges against African culture and traditional practices as causes of Africa's underdevelopment are commonplace in the Western scholarly and mass media. Africa's perennial problem of bureaucratic corruption is said to be rooted in the practice of gift-giving, which is of great antiquity throughout the continent. Meanwhile, administrative ineptitude is said to be a function of Africa's lack of experience with statecraft. Africa's economic problems are believed to result from the absence of a history of economic development throughout the continent. Furthermore, traditional Africans are said to be risk averse, and possess no notion of the meaning or essence of economic profit. As for the continent's present technological condition, it is said to result from a lack of experience with technological innovation. A related but more outlandish claim is that Africans have never contributed anything to human civilization. The myth of Africa as a dark continent has its roots in these and similarly erroneous but popular claims.

My desire to interrogate such claims as well as theories that trivialize the accomplishments of Africans and denigrate their culture and traditional practices was recently rekindled by questions from students in two of my courses, 'Africa in World Affairs' and 'Urban Planning and Development.' The students are always flabbergasted by accounts to the effect that Africans had made remarkable accomplishments in domains such as architecture and building construction, spatial organization, governance and public administration, politics and religion prior to the European conquest. On my part, I was, and continue to be, disheartened by the popularity, in the West, of the myth of Africa as a dark continent. This work constitutes my modest contribution to efforts aimed at effacing this myth. The project seeks to accomplish the following specific goals:

- demonstrate that the myth of Africa as a dark continent prior to the European conquest was simply that – a myth; and
- show that numerous indigenous African customs and traditional practices can contribute positively to development planning efforts on the continent;
- provide specific examples of areas in which indigenous traditional and culture are more likely than their Western equivalents to promote development in Africa.

Few of my students are indifferent to African customs and traditional practices; and even fewer are prepared to embrace most aspects of the culture. On this score, I must admit that the students are more tolerant of the culture than the average Westerner. For the average Westerner, every aspect of traditional African customs and beliefs is bizarre.

Ironically, Western customs and beliefs are just as weird to the traditional African. The following true story may shed some light on my assertion. A friend and colleague of mine was beginning the first week of a one-year fellowship in an African country when the news of a hunter who had just shot and killed a farmer was making the headlines in the local press. The incident occurred at night in the heart of a forest known for its rich fauna. The hunter had been arrested and was awaiting trial for murder. By the time my colleague was concluding his sojourn in Africa, he learnt that the hunter was found not guilty, discharged and acquitted. My colleague was curious and wanted to know what the hunter's defense was. According to press reports, the hunter argued that he had aimed and shot several times at a tiger. However, when he rushed to retrieve the tiger, he discovered to his dismay, the bullet-ridden remains of a human being. To bolster his claim, he had enlisted the testimonies of African traditional healers/priests, who testified to the effect that the victim was endowed with powers to transform himself into animals – something he was wont to frequently do. My colleague could not imagine how any sane jury or judge found such a 'lousy and baseless' defense persuasive.

On that same day, he began the return leg of his journey back to the United States. On the plane, an African acquaintance sitting next to him had come across an interesting story in one of the in-flight American newspapers. The story was about a man who had been found not guilty for shooting his wife. The defendant's excuse was that he had carried out the shooting while in his sleep. According to the medical experts who had served as expert witnesses, the defendant suffered from a sleeping disorder known as somnambulism. The African brought the story to my colleague's attention, expressing disbelief that such an absurd defense could be entertained by any court of law worth its salt. My colleague did not find anything absurd about the defense. However, upon reflection, he quickly realized his ethnocentric proclivity in considering as bizarre the African hunter's defense but not that of the American somnambulist.

Africans view as weird the fact that Westerners must typically leave their parents' home once they attain majority (usually 18 years of age). An African may live in the family compound, that is, the parents' home, for as long as it is necessary to do so.

Traditional Africans also find it bemusing when Westerners charge their consanguine relations for use of their property such as a home or a car. African tradition requires individuals with the means to provide all sorts of necessary assistance to their needy kin. Accordingly, it is commonplace to find a typical African financially supporting his/her nephews, nieces, cousins and other family members at all levels of education. The beneficiaries are never required to repay for such support. Rather, they are expected to assist other needy family members once they are capable of doing so. In the West, where 'there is no such thing as a free lunch' such expectations are absent.

The resiliency of aspects of African tradition such as these and others discussed in this book – in the sense that they have survived centuries of brutal assault – must be astonishing even to the most stoic critic. Despite their roots in pre-colonial Africa, the traditional practices examined here are of contemporary importance. Accordingly, the book explores their relevance or irrelevance to ongoing efforts to promote socio-economic progress on the continent. I spend some time exploring proposals that advocate the supplanting of indigenous cultural practices and values with Western equivalents as a viable strategy for promoting development on the continent. Furthermore, I challenge claims to the effect that Africans are inherently a 'backward people' by cataloguing their achievements in many domains. An important aim of this exercise is to provide some grist for the mills of development planners striving to improve living conditions in Africa. For far too many years, development planners have ignored African history and especially the essence of African culture and traditional practices as a foundation for socio-economic development efforts on the continent. Yet, efforts to promote development in Africa have no chance of registering significant positive results unless they are grounded in the continent's rich tradition and culture.

Acknowledgements

Extensive projects such as books inevitably involve the input of many people, with or without their knowledge. This one is certainly no exception. As stated above, the idea for this book grew out of my interaction with my students. Thus, I owe a debt of gratitude to all the students who have taken my courses on 'Africa in World Affairs' and 'Urban Planning and Development' at the University of South Florida. These students are likely to be unaware of the multi-faceted ways in which their thought-provoking questions motivated me to embark on this project. I also wish to acknowledge my indebtedness to my colleagues in the College of Arts and Sciences, particularly, David Carr, John Arthur and Kathy Weedman. David served as my 'human encyclopedia' on Western history, and Kathy's and John's profound knowledge of anthropology, and extensive experience in Africa, were very valuable to my efforts.

At a personal level, I wish to register my most sincere gratitude to the late Mama Elizabeth Njweng Njoh, my 'father,' for all her moral and other countless forms of support. It is to her that this work is dedicated. Mama, as every member of the extended Njoh family fondly referred to her, epitomized the twin concepts of 'female husband' and 'female father.' She assumed control of the Njoh estate when all of her male siblings, including my own biological father, had passed on. As the eldest surviving sibling, she was 'recalled' from her marital home – a process that was facilitated by the fact that none of her own biological children was alive. To all of the Njoh children, Mama was our 'father' and to our mothers, she was 'the husband.' She was, for all practical purposes, 'the Man.' In public, she could always be found in the company of men as opposed to that of women. She was extended every courtesy due men in traditional Cameroonian society.

I remember quite vividly that on my first day in elementary school, I gave the name, 'Mama,' when the enrollment clerk asked for my father's name. My discussion of the African family (Chapter Four) and 'women, sexuality and property inheritance' (Chapter Six) is informed by my first-hand knowledge of life in the Njoh family, an African family that was headed by a 'female father/husband.'

I also benefited in untold ways from exchanges with my mother, Mrs. Eni Diana Njoh. She is an avowed, but certainly not a 'right-wing' or 'fundamentalist' Christian. She was quick to comprehend the imbecility of requiring Africans to assume Western names upon baptism or conversion to Christianity. She was supportive of my decision to abandon the Western name she had given me when I was baptized as an infant. More importantly, she was persuaded by my argument to the effect that any name, be it of African, European or any other origin, could serve as someone's Christian name. As an avid proponent of free-thinking, my mother has been instrumental in my intellectual growth and development. For this, and other myriad reasons, I thank her from the bottom of my heart.

I would be remiss if I failed to express my hearty appreciations to all other members of the immediate and extended Njoh family, especially the Njoh household in Tampa, Florida, USA, including Fri, Eni, Fon, and Akwi. I heartily thank my uncle, Ni Fomunyoh Azobo. His impressive knowledge of Meta culture and life in traditional Meta society was invaluable to the realization of this project. The word 'uncles' in Meta language refers exclusively to one's maternal uncles. One's paternal uncles are referred to as one's fathers.

Finally, I thank all the people – and there are too many to name – who were willing or unwilling participants in my endless discussions and debates of the merits of African culture and traditional practices. While I remain eternally grateful to everyone who contributed in any manner, shape or form to the realization of this project, I take full responsibility for all errors of omission or commission contained therein.

Ambe J. Njoh,
University of South Florida St. Petersburg, Florida, USA.

Chapter 1

The Culture–Development Nexus

Introduction

Africa continues to lag behind all regions of the world on every established indicator of development. This point is hardly contentious. What constitutes the subject of fierce and often, rancorous debate is the question of why the continent remains perpetually underdeveloped despite national and international efforts to reverse the nightmarish situation. This question is not only germane, but also, its centrality in the discourse on development is undeniable. Yet this book is not about the causes of Africa's underdevelopment *per se*. Rather, it is about how African customary and traditional practices, and the efforts to obliterate these practices by agents of Western civilization affect, or can potentially affect, socio-economic development in Africa. The book has two secondary objectives. The first is to contribute to efforts seeking to dispel the myth of pre-colonial Africa as a 'dark continent.' It accomplishes this objective by highlighting the accomplishments of Africans in domains such as family and social welfare, gender relations, health care, public administration, architecture and housing, prior to the European conquest. The second is to interrogate arguments that advocate supplanting African culture and tradition with Euro-centric values as a panacea for Africa's development quandary.

The link between culture and development

At the heart of one strand of the debate surrounding Africa's development/ underdevelopment is African culture. I employ this obstinate and nebulous term here, and throughout this book, to refer to the knowledge, beliefs, customs, morals, tradition and habits of African people. The implied link between culture and development is itself debatable. In its policy report, 'Our Creative Diversity,' The World Commission on Culture and Development (WCCD)[1] (hereafter, the Commission), questions the wisdom of inferring a link between culture and development. The report, written under the leadership of former UN Secretary General, Javier Perez de Cuellar, opines that it is meaningless to focalize on this nebulous relationship given especially that

---

1   The World Commission on Culture and Development (WCCD) was jointly constituted by the United Nations and the United Nations Educational, Scientific and Cultural Organization (UNESCO) in October 1992. Its main charge was to study and prepare a policy-oriented report on the interactions between culture and development. It completed and submitted the said report in November 1995.

development is part of a people's culture. The report further laments what it perceives as ambiguities arising from the ideological question of whether 'culture' is at once the aim and end of 'development' (WCCD, 1995).

De Cuellar and his colleagues commit a number of common errors. First, they fail to decouple the notion of development. In this regard, they define development 'as a process that enhances the effective freedom of the people involved to pursue whatever they have reason to value' (WCCD, 1995: 1). Second, they ignore the essence of development as a universal concept. In fact, by treating culture as an element, rather than an instrument, of development, they imply that some cultures may be oblivious, while others may be fully aware of, and attentive to, development since not all elements constitute common denominators in every culture. Nothing could be further from the truth. To be sure, every culture the world over has some concept of development and aspires to attain it. The difference surfaces only when the concept is decoupled and deconstructed. Such an exercise may reveal that the notion of what constitutes development in one culture may differ dramatically from, and in some cases conflict sharply with, what comprises development in another. For example, Euro-centric culture, in which the capitalist ideology and Protestant ethic are rooted, places much premium on capital accumulation, entrepreneurial attitudes, and material wealth *inter alia*. As we shall see, while these attributes are not completely absent, they are certainly not at the top of the priority ladder, in the context of traditional Africa.

What the foregoing narrative suggests is that while development may be a universal objective, as a concept, development lays no claim to a universal definition. To be sure, the concept assumes a bewildering variety of meanings. I have no intention of either regurgitating or analyzing the various meanings that are often associated with the concept of development here. However, it is necessary to re-examine the Commission's definition stated above. This definition, the Commission claims, deviates significantly from orthodox conceptualization schemes particularly because of its sensitivity to varying cultural values.

However, the rather tenuous definition implies that there is hardly any human activity that does not qualify as development. All it takes to be viewed as development is for the activity to be conducted by 'a people,' who have reason to value such an activity. Yet, there is a plethora of human activity and actions such as racial and gender-based discrimination, human rights abuse, bribery and corruption, which although some 'people have reason to value,' can by no means be regarded as development. Certain aspects of cultural and traditional practices in Africa in particular and other parts of the world in general, encourage behavior that constitutes impediments to development, however it may be defined. The question is: which are these aspects? This question is at the heart of the discussion in this book.

The incrimination of African culture as a leading cause of (socio-economic) 'backwardness' led European colonial and imperial authorities to take a number of dubious actions in Africa. For instance, to facilitate attainment of the economic objectives of the colonial enterprise, European colonial authorities and their collaborators proceeded with unparalleled gusto to institute policies designed specifically to supplant African culture and traditional practices with Euro-centric varieties. Thus, colonial authorities and other disciples of Western civilization believe that Euro-centric values are more likely than their African equivalents to facilitate the attainment of development objectives in Africa. This book challenges this belief, which has always been central to colonial and post-colonial development initiatives in Africa. The book essentially contends that the Eurocentric values, customs and traditions that have been adopted in Africa since the colonial era have hurt, rather than helped the development process. African traditional practices and culture hold more promise than Euro-centric equivalents in many domains, including but not limited to, primary healthcare, the family, gender relations, community development, housing, resource mobilization, self-help development initiatives, and local administration.

The civilizing mission of European colonialism

Efforts along these lines constituted part of a meticulous plan to 'civilize,' or to state it more accurately, 'Europeanize,' Africans. The concern with African culture as a hurdle to European colonial development objectives led some European powers such as France to redefine and couch their colonial goals in terms of a mission to civilize *(or la mission civilisatrice)* what they viewed as the 'backward' and 'primitive' people of Africa. It is therefore hardly any wonder that the French were pre-occupied with acculturating and assimilating Africans. Having said that, it is important to state for the record, that there is a penchant, on the part of analysts of European colonialism in Africa, to associate acculturation and assimilation as objectives of the colonial enterprise, exclusively with the French. Yet, colonial powers, without exception, were bent on assimilating and acculturating Africans. The results were, and continue to be, clashes or crises of cultures, or what I herein refer to as 'acculturation crises.' The principal objective of acculturation and assimilation efforts on the part of colonial powers was to transform Africans into 'Black Europeans.' Part of these elaborate efforts entailed a heavy dose of ethnocentricity on the part of Europeans and propaganda campaigns designed to deride African traditional practices while exalting European values.

The European mission to 'civilize' Africans met with fierce resistance on the ground. Most Africans were unimpressed by European culture and clung tightly to their traditional practices and institutions. In response, colonial authorities proceeded to employ a number of measures, including force, to inculcate European values in what they considered to be 'barbarians.' In most cases, a carrot-and-stick approach was employed. Thus, for instance, the French labeled Africans who had subjected themselves to French or European acculturation and assimilation as *'les evolués'* and accorded them preferential treatment in society.[2] Those who insisted on maintaining

---

2 The term *'evolués'* was designed to suggest that the assimilated and acculturated Africans had 'evolved' to a level on the modernization ladder that was deemed superior to that occupied by the rest of the natives.

their 'Africanness' were subjected to discrimination and exclusion from colonial social, economic and political activities. The assimilated and acculturated individuals had the privilege of living in European-style housing located in highly sought-after areas, close to the European Quarters. Additionally, they were given employment in the colonial civil service, and enjoyed the attendant social and economic benefits.

Christianity and the erosion of African culture

Christian missionaries, who worked alongside colonial authorities, were also active in the acculturation and assimilation process. They had moved very early during the colonial era to spread the gospel throughout Africa. Prominent amongst the many hurdles they encountered were African traditional practices, which they viewed as antithetical to Christian doctrines. If Christian missionaries viewed African culture as a hurdle, they hardly considered it to be insurmountable. From the following utterance by one missionary authority, it is arguable that African culture was seen as constituting no more than a menace that could/can be eradicated with facility.

> The Army of the Cause, advancing at the bidding of the Lord to conquer the hearts of men, can never be defeated, but its rate of advance can be slowed down by acts of unwisdom and ignorance on the part of its supporters (Universal House of Justice, On-Line).

Hence, as their colonial counterparts, missionaries were always bent on supplanting African spiritualism and belief systems with Eurocentric Christian versions. Christianity entailed baptism, which went along with accepting a Eurocentric Christian name. This invariably supplanted the new Christian's African name. Such was an important element in the acculturation process – it was symptomatic of 'modernity.' Christians and their families had the privilege of receiving 'modern' medical care and their children could attend mission schools. These were but a few of the benefits reserved for Africans who decided in favor of separating themselves from their traditional reality and existence. Those who resisted this transformation were pejoratively labeled as 'pagans' or 'animists' and denied any socio-economic benefits that were extended to Christians.

We will return to these issues a little later in this chapter and will delve into them in more detail in the chapters that follow. For the moment, suffice to state that efforts to eradicate African traditional practices were never preceded by any meaningful attempt to determine their actual negative and positive attributes. The rationale for their elimination was simply predicated on the fact that they were of African origin and anything African was invariably inferior and worthless in the eyes of Europeans. The only exceptions were the continent's natural resources, which Europeans were determined to exploit at all cost.

Development economists and cultural erosion

At the eve of independence for most African countries in the 1950s, development economists and international development agencies were beginning to seriously contemplate the necessary strategies for facilitating development in the emerging nations. Again, the view that African culture was antithetical to socio-economic development re-asserted its dominance. The concept of development then was narrowly defined in exclusively economic terms. Thus, development was seen as constituting a sustained rise in gross national product (GNP).

Leading development economists such as Walt Rostow suggested that it was impossible for Africa to develop without abandoning its traditional practices and assuming Euro-centric cultural values, beliefs and ideology. Rostow summarized his views about the prerequisites for economic development, which entailed, as an absolute necessity, cultural transformation, in *The Stages of Economic Growth: A Non-Communist Manifesto* (1960). This work instantly became the Gospel of development, as one development agent after another went to work inculcating in Africans, Euro-centric cultural values, ideologies and philosophies. According to Rostow (1960), societies such as Africa were at the first of five stages that all societies must pass through in order to become developed. He referred to societies at this first stage as 'traditional societies.' The other stages include, 'the pre-condition for take-off,' 'the take-off,' 'age of high mass-consumption,' and 'beyond consumption.'

The notion of traditional society

The term traditional society, clearly defined in the minds of the development economists of that time the continent of Africa, which many had come to see as the 'Dark Continent.' However, to be fair to Rostow, he grouped under this rubric the entire pre-Newtonian world, including Medieval Europe, the civilization of the Middle East, and the dynasties of China. The traditional society is known for its adherence to what Westerners consider 'primitive values,' with agriculture constituting its most dominant economic activity. Here, the extended family, village and clan play a predominant role in social organization. It is this organizational unit that owns land and other valuable entities in society. Members of traditional societies are also said to be risk-averse, remain oblivious to their environment and lack the ability to manipulate this environment to their economic advantage.

Believing that there was a 'one best way' to development, development economists, especially in the late-1950s and early-1960s argued that in order to develop, developing countries had to follow the same path that the contemporary developed countries had followed. In Africa, as stated earlier, the most significant impediment to this process was identified as the continent's cultural practices and traditional institutions.

Re-conceptualizing development

As the 1960s drew to a close, some dissenting voices could be heard in the development economic community. These voices, prominent amongst which was that of Dudley Seers, began to question the sagacity of defining the concept development in strictly economic terms. In a seminal article captioned 'The Meaning of Development,' which appeared in *The International Development Review*, the official outlet of the Society for International Development, Seers made a persuasive case against defining the concept of development in exclusively economic terms (Seers, 1969). Such a definition, Seers argued, tended to ignore other important attributes of development, such as levels of unemployment, levels of socio-economic inequalities, levels of poverty, and other conditions to which citizens of any given country may be subjected. Eight years following the publication of this work, Seers re-visited the issue of defining development (see Seers, 1977). Conceding that his critique of the conceptualization of development in exclusively economic terms left something to be desired, he added yet another important indicator of development, namely level of dependence. Thus, according to Seers, a polity cannot be considered developed whilst it continues to depend on others to meet the basic needs of its people (Seers, 1977).

The concept of dependence as employed by Seers is broad enough to include anything from cultural to economic dependence. In this regard, it will be foolhardy to suggest that African countries adopt the development model that was adopted by European and other contemporary developed countries. To appreciate this, it is necessary to understand the notion of culture as encompassing a total way of life, which embraces how a people treat death, how they greet or receive a newborn, what they eat, what they wear – in fact, how they live (Rodney, 1981). Efforts to alter the taste and consumption habits of Africans, which were central to the assimilation and acculturation initiatives of imperial powers and other agents of Western civilization, went a good way to ensure the continent's socio-economic and technological dependence on Europe in particular and the Western World in general.

Efforts to deconstruct the culture and traditional practices with a view to addressing the vital questions regarding the relationship between culture and development have been scant at best. This is particularly true in the case of Africa. This rationalizes, to a small degree, the view that African culture was antithetical to development, which held sway during the colonial era, the immediate post-colonial period and continues to be popular within some circles today. However, it is necessary to state that this view is largely unsubstantiated by empirical evidence.

African culture, an impediment to development?

The difference between African and Euro-centric culture alluded to earlier is arguably at the root of the view on the part of colonial powers and later, some development economists (e.g., Rostow, 1960; Bauer, 1972) that African culture is antithetical to development. For a long time, as mentioned above, this view reigned supreme and was seldom debated or challenged. Here, it is important to draw attention to the fact that the seemingly elusive concept of development was defined narrowly to include economic growth engendered by a rapid and sustained expansion of production, productivity and income per head. Even if we disregard the limitations of this narrow definition, can we meaningfully argue that all aspects of African culture and traditional practices are antithetical to development?

As much as I am unimpressed by any argument that indiscriminately incriminates all aspects of African culture as impediments to development, I am equally not persuaded by arguments that exalt every thread of this culture as a real or potential facilitator of development. Elsewhere, I have cautioned against the dangers of being beguiled by sentimentalism when discussing African culture and traditional practices 'as such can be justifiably subjected to the same criticisms as its converse, ill-founded disparagement' (Njoh, 1999: 31). More recent attempts at analyzing and refining the concept of development suffer from the same deficiencies as those of the late-1950s and 1960s. They have paid only passing attention to the link between culture and development. More importantly, hardly any effort has been made to determine the potency of Eurocentric culture and values vis-à-vis the African varieties they usually seek to supplant, as instruments of development. Consequently, when faced with the need to attain a development goal, planners and other agents of development tend to seek Western means while ignoring considerably more potent indigenous African alternatives.

Objective, focus and central questions of the book

An important objective of this book is to contribute to efforts addressed to reversing this tendency. The book differs in many ways from previous works that capriciously condemn (e.g., Rostow, 1960; Bauer, 1972), or indiscriminately endorse (e.g., Rodney, 1981; Jarrett, 1996), all aspects of African culture and traditional practices. First, its thematic focus is on the link between culture and development. Although the importance of this link is widely acknowledged, efforts to systematically study it remain woefully inadequate. On the one hand, there are unsubstantiated claims to the effect that African culture constitutes an impediment to socio-economic development (e.g., Rostow, 1960; Bauer, 1972). Accordingly, initiatives to eradicate and replace this culture with Euro-centric values were designed to foster development in Africa. On the other hand, some hold that efforts to supplant African culture with Euro-centric varieties have helped to under-develop Africa (e.g., Jarrett, 1996; Rodney, 1981). In either case, it is not clear which aspect of African culture (e.g., arts, communication, language, technology or the fundamental belief systems that frame behavioral and social patterns in Africa) is being extolled or castigated.

For another thing, and in an effort to help dispel the myth of Africa as a 'dark continent' prior to the European conquest, it identifies and discusses a number of major accomplishments of pre-colonial Africans. Furthermore, it analyzes salient attributes of major cultural and traditional values, beliefs and practices from across

Africa with a view to showing how they may constitute impediments or facilitators to the continent's development aspirations. Finally, its regional focus is limited to sub-Saharan Africa, especially because of the similarities in terms of culture, politico-economic and social development of countries in the region. Perhaps more importantly for the purpose of the discussion in this book, Rodney (1981) has drawn attention to the fact that sub-Saharan Africa constitutes a broad community of nations with clear and easily discernible resemblances. This regional focus does not preclude the culling of examples or illustrations from the region north of the Sahara.

Plan of the book

The endeavour to uncover answers to these questions progresses through the eleven chapters summarized below. To provide a backdrop for the discussion in the rest of the book, Chapter Two discusses pre-colonial Africa. By discussing pre-colonial Africa, the chapter hopes to avoid the common blunder of treating African culture as static and failing to account for the developments and achievements of early African civilizations in analyses of the continent's present condition. Accordingly the chapter seeks to demonstrate the dynamism of African culture and underscore the achievements of Africans in different walks of life prior to the European conquest that occurred in the late-19th century. Particular attention is paid to the tradition, culture, belief systems, and achievements in the areas of trade, craft, health care, political organization and governance of Africans. In the area of technology, for instance, it is important to note that Africans had developed proficiency in working with iron as far back as 200 A.D. – that is, long before the iron age in Europe.

Chapter Three focuses on the advent of colonialism and Christianity, and how these two inextricably intertwined enterprises worked to supplant African traditional culture and value systems with Euro-centric varieties. Of critical importance in the chapter are the strategies that these two imperial agents employed as well as the overt and covert reasons that were advanced for their indefatigable efforts to transform African society. The chapter is certainly not oblivious to the role played by Arabs with the introduction of Islam in transforming the society a good while pre-dating the arrival of colonial powers and Christian missionaries on the continent. However, the chapter pays only passing attention to Arabic/Islamic influences, particularly because it can be persuasively argued that Euro-centric values and ideology – of which Christian doctrines must be considered a part – have done exceedingly more than all other factors to shape socio-economic conditions in Africa.

In Chapter Four, the attention is turned to what is arguably the most important and most resilient institution in African society, the family. In the highly acclaimed video series, *The Africans*, eminent Africanist, Ali Mazrui, impresses upon viewers that the African family has surprisingly survived unscathed, centuries of assault from exogenous forces. Unlike other institutions in Africa, the African family remains a unique phenomenon. It is extended, not only in the rural areas, but also in the urban centres and the diaspora. A barrage of incessant propaganda campaigns on the part of missionaries and agents of Western civilization has done little to transform the African family into a clone of the Western family. Africans continue to practice polygamy and have steadfastly resisted the temptation to view the family through a Western prism. Thus, for instance, Africans have no equivalence for English words such as cousin, niece, nephew, and paternal uncle. The brother of one's father is also referred to as one's 'father,' while one's nephews and nieces are considered one's children, and one's cousins (X times removed) are referred to as one's brothers and sisters. Upon the death of a married man, a designated brother or any other designated kin (in the absence of an adult male sibling) assumes responsibility of the man's wife, children and property. The chapter discusses these and cognate phenomena, and explores their implications for development on the continent.

Chapter Five examines the relationship between people and land in traditional Africa. The importance of land in the political economy of Africa has been accentuated in recent years by developments in Zimbabwe. The overly simplistic and sensationalized accounts in the European and American popular media justly or unjustly vilify the country's President, Robert Mugabe. These accounts hold that Mugabe's decision to confiscate, and transfer large tracts of previously White-owned land to members of the country's Black native population was predicated on reasons of political expediency. Woefully absent from these accounts is an acknowledgement of the vast gulf separating the African and Euro-centric models of land 'ownership.' The Euro-centric model – later imposed on Africans by colonial authorities – is based on the theory that whoever possesses economic and/or political power also has power over land. This philosophy treats land as any other commodity that can be bought and sold on the market. The African model differs sharply and in fact, conflicts with the Euro-centric model, particularly on account of the fact that it does not treat land as a commodity. More worthy of note is the fact that African customary laws recognize only groups of people or communities, and certainly not individuals, as the entities that may control or 'own' land. An African chief once painted a lucid picture of the traditional African view of the concept of land ownership, which unequivocally contradicts the Euro-centric model, in the following statement (Quoted in Meek, 1949: Title Page).

> I conceive that land belongs to a vast family of which many are dead, few are living and countless members are still unborn.

The chapter probes the distinguishing characteristics of the two models. Additionally, it highlights the problems resulting from the introduction of European land tenure models and the concomitant commodification of land in Africa. Of great importance in the process are the actual steps that colonial officials and the indigenous leadership after them took to supplant traditional African land tenure systems with Euro-centric varieties. Finally, the chapter deals with a number of 'what if' questions. For instance, 'what if' European colonial authorities did not move to replace traditional African land tenure systems with European varieties? In other words, what are the plausible impacts of traditional African land tenure systems – including African beliefs about the value and role of land in people's lives – on contemporary development efforts on the continent?

Chapter Six explores 'women, sexuality and property inheritance.' This topic encompasses issues of critical importance in the discourse on development, such as women's roles in production, reproduction, marriage, domestic labour, health, politics and societal governance. These roles, which have always defined the status of African women, have changed significantly since the colonial era. It is generally agreed that the changes have meant a rapid erosion of the traditional culture and values that delineate the status of women and gender relations in Africa. What is however, debatable is the effect of the rapid erosion of culture and traditional practices on the status of African women. Amongst other things, the chapter examines the nature and purpose of two such practices, female circumcision – or what is often inflammatorily referred to as female genital mutilation (FGM) – and polygamy. These practices constitute the subject of heated debate in the discourse on the status of women in African society.

Opponents of African culture and traditional practices, such as Western feminists, contend that these practices contribute to diminishing the status of African women. Consequently, their efforts to improve the status of African women have included campaigns to eradicate these practices, especially female circumcision. In their bid to eradicate polygamy in Africa, Western feminists have always had an unlikely, but dependable ally in Christian religious authorities. Christianity implored Africans to forsake their traditional practices and taught them that polygamy is ungodly. This, of course, is despite the pervasiveness of polygamy in biblical narratives. Thus, the chapter questions the motives of efforts to replace African traditional practices on the part of Christian religious authorities and other agents of Western civilization in Africa.

Proponents contend that the erosion of African culture and traditional values have had negative implications for the status of African women. For instance, women dominated the prestigious agricultural sector and enjoyed the attendant status during the pre-colonial era. Women lost this prestigious status during the colonial era (Strobel, 1982; Boserup, 1970). The gender-based division of labour schemes crafted and propagated by colonial authorities assigned women to the subsistence or food crop farming sector while men were charged with the task of cultivating cash crops destined for the metropolitan economies. Consequently, women earned far less than their male counterparts for the same amount of labour input since cash crops fetched a lot more than food crops on the market. In the documentary series, *The Africans*, Ali Mazrui advances an argument of relevance here, along the following lines. Efforts that have encouraged women to abandon rural areas and traditional roles such as farming in favor of low-level clerical activities in urban areas in the name of 'modernity' have contributed to effectively diminish the status of women in Africa. Ifi Amadiume locates the blame for the erosion of African values in the sphere of gender relations squarely on the shoulders of colonial authorities when she advances the following argument (Amadiume, On-Line).

The colonial experience that introduced Western gender perceptions and practices affected the traditional involvement of African women in the development of their societies, leading to women's marginalization and economic and political disempowerment.

How accurate are these claims and counter-claims? Have changes culminating in the erosion of traditional African values and the attendant replacement of these values with Euro-centric varieties helped or hurt the status of women in African society? These questions are at the heart of the analysis in this chapter.

Chapter Seven explores traditional African political and administrative philosophy and practice. Contrary to popular opinion, the existence of sophisticated politico-administrative systems pre-dated the arrival of European colonial powers in Africa in the late-19th century. These systems were, for the most part, based on democratic principles – again, contrary to popular accounts, which portray the African king as despotic and autocratic. Certainly, monarchies were common in pre-colonial Africa, with the kings and queens having as a major responsibility, ensuring stability and order within the territories under their jurisdictions. Above all, the kings/queens were responsible for defending their territories against invasions. However, a fact that has been systematically omitted in the relevant literature is that a king/queen did not govern single-handedly and could be asked to relinquish power if he/she did not rule in the best interest of the people. A council of elders, the chief priests, a group of moral elders and chiefs assisted the traditional African king/queen in the execution of his/her official functions.

The chapter delves further into the composition and functioning of the traditional politico-administrative system. For now, it is important to note that colonial authorities, save the British, worked fervently to supplant the traditional politico-administrative institutions. The French, who are often identified with the colonial strategy that came to be known as 'direct rule,' strived to replace the traditional systems with French politico-administrative institutions. On their part, the British, who are generally credited with popularizing the 'indirect rule' strategy, maintained and implemented their colonial development policies through the existing traditional African institutions. Here, I hasten to note that the British did so not because they were interested in safeguarding or preserving African tradition and culture but particularly for reasons of political expediency. Working through extant institutions was convenient, efficient and served to minimize opposition to colonial rule. The French might have failed to appreciate this because of their infatuation with the acculturation and assimilation objectives of the colonial enterprise.

However, I would be remiss if I failed to acknowledge the fact that as the demise of the colonial era in most of Africa drew near in the 1950s, the British decided in favor of abandoning the 'indirect rule' model (Crook, 1986). Thus, it is safe to state that colonial authorities, without exception, contributed to the destruction of African traditional politico-administrative systems. How and why did they do this? Which traits of these systems hold promise for ongoing and future efforts to improve governance in African countries? Which of the traits have served to impede such efforts? The chapter is preoccupied with these and cognate questions.

Chapter Eight focuses on traditional African resource mobilization strategies. Territorial governance, which as mentioned above, was one of the major responsibilities of kings/queens in pre-colonial Africa, could have hardly found expression in reality without public infrastructure. Public infrastructure particularly roads, were as necessary to broadcast authority then as it is in any polity today (Herbst, 2000). To attain this objective, kings/queens had to develop an impressive inventory of roads. For instance, the Ashanti King, whose empire is said to have possessed some of the characteristics of a modern nation-state, developed an elaborate system of roads that converged on the empire's capital, Kumassi (Herbst, 2000). Certainly, the development of such elaborate infrastructure could not have been possible without a significant pool of resources. It is easy to appreciate how pre-colonial Africans would have been successful in completing such large-scale infrastructure projects once we understand the essence of communalism, which is a key attribute of the traditional African value system. As I have observed elsewhere, (Njoh, 2003b), it was common practice in pre-colonial Africa for members of enclaves, towns and villages to come together for the purpose of executing local development projects. Thus, people considered it a duty to participate in long-term community development activities such as the building of roads, market places, the king's palace or any other communally owned piece of infrastructure. Also, it was commonplace for pre-colonial Africans to indulge in short-term activities such as the hunting of game for communal consumption. At the micro-level, pre-colonial Africans had also developed sophisticated project development mechanisms. For instance, it was common for individuals to organize themselves into small manageable groups for the purpose of executing tasks that would otherwise be daunting to an individual working solo.

In this case, a group consisting of say, six persons, A, B, C, D, E, and F, may congregate as a team to complete the task of each member of the team on a rotating basis. Thus, if they began working for A on Monday, they will jointly work for B on Tuesday, and so on, and will end a one-week work cycle by working for F on Saturday. I refer to such groups as rotating task associations (RTAs). When their purpose is to work as groups to raise funds for individual members, they are commonly referred to as rotating savings and credit associations (ROSCAs). If the same group of six mentioned above was a ROSCA, they could agree to contribute say, $10.00 each, for a total of $60.00 and give to A on Monday, and do the same for B on Tuesday and so on, and end the cycle by contributing and giving F the same amount on Saturday.

Such groups qualify as what goes under the appellation, mutual self-help groups in Europe and North America. However, it is important to note that an essential component of the mutual self-help ethos in traditional Africa calls for members of mutual-help groups not only to receive, but also to give help. Thus, each member of a mutual-aid group is expected to assume the role of a help-beneficiary ('helpee') and that of a help-giver (helper). This sets mutual-aid groups in Africa apart from identical groups in Europe and North America, where individuals join self-help groups, such as Alcohol Anonymous (AA), to receive, rather than give, help. This chapter examines the philosophy of communalism and the mutual self-help ethos in traditional or pre-colonial Africa. One aim of this exercise is to incite thinking on how Africa's past can serve as a repertoire of knowledge for ongoing efforts to promote community participation as a strategy for implementing labour- and capital-intensive projects on the continent in particular and in other resource-scarce locales in general.

Chapter Nine examines traditional African public health and healing strategies. Prior to the arrival of European colonial powers, Africans had developed health care strategies that were adapted to, and defined by, their culture, beliefs and natural environment. These strategies, which contain a heavy dose of 'herbalism' and 'spiritualism,' were not acquired through formal training but extensive apprenticeship. Arguably, this constitutes one reason why Westerners are wont to reject or ignore traditional African medicine and healing strategies. The Western popular media frequently ridicules traditional African healing practices, pejoratively referring to such practices as 'voodoo' or 'black magic.' Christian religious authorities, dating back to the colonial era have worked relentlessly to discourage such practices, which they consider 'ungodly' and sinful. The actions of these and cognate authorities vis-à-vis African traditional healing practices is a function of, amongst many other things, ignorance. The chapter contributes to efforts designed to dispel this ignorance. To be sure, such efforts are ongoing. For instance, the World Health Organization (WHO), now recognizes the potency of traditional medicine. According to the WHO, traditional medicine includes (WHO, 2002: 7),

> Diverse health practices, approaches, knowledge and beliefs incorporating plant, animal and/or mineral based medicines, spiritual therapies, manual techniques and exercises applied singularly or in combination to maintain well-being, as well as to treat, diagnose or prevent illness.

As further evidence of the growing recognition of the viability of traditional healing methods, an international conference under the title 'African Healing Wisdom: From Tradition to Current Application & Research,' took place at the Hyatt Regency Hotel in Washington, DC from 6–9 July 2005. This chapter seeks to accentuate the need for co-operation between African traditional healers and modern biomedical scientists as a viable strategy for combating new and re-emerging diseases in Africa in particular and the world at large.

Chapter Ten is pre-occupied with 'traditional architecture and housing strategies.' The relevant literature is replete with disparaging accounts of the housing of pre-colonial Africans and contemptuous characterizations of their architecture as lacking 'a feeling of space as we understand it' (see e.g., Prussin, 1986: 3). These accounts lead one to believe that traditional or pre-colonial Africa was devoid of architectural and physical design structures worthy of note. Yet, nothing could be further from the truth. This chapter marshals evidence to show that contrary to popular belief, traditional Africans were particularly adept at manipulating habitable space. Thus, there exists a rich array of architectural and housing schemes that can be considered purely African by origin. These pre-colonial or traditional African schemes are particularly unique because, as Denyer (1978: 4) notes, they constituted 'a personal adaptation of a group solution.' The buildings and other structures developed by any given society 'were in a style which had been communally worked over several generations and consequently were closely tailored to the needs of its people' (Ibid, p. 4). The chapter provides evidence to support these and cognate assertions. Ultimately, the chapter argues that in contrast to the Euro-centric architecture and housing schemes by colonial authorities the traditional varieties are more in line with Africa's natural, social, political and cultural environment. Thus, African traditional schemes are more likely than the Euro-centric varieties to facilitate attainment of the development objectives of African countries.

As the concluding chapter, Chapter Eleven draws from evidence presented throughout the book to challenge the popularly-held view that the fate of Africa is sealed by the nature of African culture and tradition. The chapter views indigenous culture and tradition as an asset rather than a liability to development efforts on the continent. In fact, indigenous culture and traditional practices are seen as the main pillars on which any sound development plan in Africa must repose. Finally, the chapter identifies and discusses means by which African culture and traditional practices can be incorporated into strategies geared towards promoting cultural and ideological goals, enhancing ecological management efforts, and facilitating regional socio-economic and political integration.

Importance and target audience of the book

This book addresses the following important, but largely ignored, question in the discourse on African development. What is the link between culture/tradition and socio-economic development in Africa? Culture, tradition, and development are seldom discussed as the inextricably intertwined concepts that they are in the context of Africa. By weaving a common thread through these concepts, the book breaks new ground in the discourse on development (in Africa).

The book will prove invaluable as the main or companion text for graduate or upper division undergraduate courses in multicultural studies, African history, politics of Africa or developing areas, colonial history, cultural anthropology, race relations, international development planning, human or development geography, and international studies. It contains material of great utility to multinational corporations, members of the academic, professional and diplomatic communities working on projects requiring extensive interaction with Africans, or those involved in natural resource exploitation, research, education and training, conflict resolution, community, regional and national development in African countries. It promises to provide the scholarly and international development establishment fodder for finding solutions to Africa's development impasse. The book can also serve as reference material for students of African political economy, culture and/or development.

Finally, and especially because it is free of technical jargon, the book can constitute a non-sensationalized source of information on Africa, Africans and the African condition for members of the general public.

Chapter 2

African Accomplishments before the European Encounter

Introduction

The term tradition has its roots in the Latin word, *traditio*, meaning to 'hand down.' Thus, we employ the term here to connote customs, beliefs and opinions that have been handed down from generation to generation. The eloquent Africanist, Kwame Gyekye (1997) defines tradition to include the 'values, practices, outlooks, and institutions' that one generation within any given society inherits from the generation(s) that preceded it. Gyekye further attempts to distinguish between the terms tradition and culture. He sees tradition as comprising a set of cultural practices that have survived through several generations. In other words, the distinction between the two concepts is duration, although he does not specify how many generations a cultural practice would have to survive to become a tradition. I use both terms in the context of the present discussion interchangeably.

Theories incriminating culture or tradition as a cause of socio-economic underdevelopment in Africa usually make two major implications. The first is that the customs, beliefs, socio-political institutions, economic systems and technology that Africans have inherited from their ancestors are antithetical to contemporary socio-economic development aspirations. The second is more cynical and goes something like this: past generations of Africans invented/created nothing of value and therefore bequeathed the present generation with no sound foundation for socio-economic development.

The following excerpt from a popular introductory history textbook by Arnold Toynbee (1946: 234) that has been widely used in the West since the end of the Second World War summarizes the major tenets of the second argument:

> The Black race has not helped to create civilization, while the Polynesian white has helped to create one civilization, the brown race, two, the yellow race, three, the red race and the Nordic white, four apiece, the Alpine white race, nine and the Mediterranean white race, ten (Toynbee, 1946: 234).

The present chapter has two main objectives, namely to debunk the theory that the 'Black race' has not contributed to civilization, and show that indigenous African customs, beliefs and institutions are not necessarily antithetical to contemporary socio-economic development aspirations. I accomplish both objectives in this and subsequent chapters by marshalling evidence to show that rather than inhibit, African indigenous customs, beliefs and institutions helped ancient Africans to make tremendous strides in the areas of commerce, trade, metallurgy, transportation, craft, education, public and local government administration, architecture, and road building.

I dedicate most of my efforts in subsequent chapters to demonstrating that Africa's contemporary problems of underdevelopment are not a product of flaws in Africa's culture and/tradition. Rather, the problems are a product of centuries of brutal transformations, aggressive and dehumanising interventions and exploitation that have been visited upon the continent by Westerners during the trans-Atlantic slave trade, the European colonial/imperial époque and the contemporary neo-colonial era. I begin by briefly reviewing the conceptual link between tradition and development.

Tradition and development

The concept of tradition commonly implies a point on the development continuum which societies must abandon as they strive to 'modernize.' Development economists, sociologists, and anthropologists who believe that the progression to modernity is a simple linear process have propounded this ambiguous view of tradition. The view constitutes the corpus of Rostow's model of modernization (see Chapter One), which embodies five specific stages, beginning with the 'traditional society' and culminating in a technologically and economically sophisticated or 'modern' society. Based on the Rostowian model, tradition and modernity are polar opposites. African countries epitomize traditional societies while Western countries exemplify the modern ones. Within the framework of this model, endogenous as opposed to exogenous factors conspire to thwart development efforts in Africa. Sorenson (2003) has eloquently summarized the major tenets of the popular theory that underdevelopment in Third World societies such as Africa is due to internal as opposed to external factors.

Basically, the theory holds that so-called traditional societies such as those in Africa are underdeveloped because of the lack of important propellants of development, including a work ethic, morals, innovative and entrepreneurial capacity, free market mechanisms, a propensity for taking risks, and organizational acumen. The absence of these factors, according to the theory, is itself a function of flaws in the culture, customs and social mores of traditional societies. Particularly noteworthy in this latter respect, is the fact that the theory considers the leading cause of underdevelopment in so-called traditional societies as the fact that such societies tend to place a lot of emphasis on kinship and family than on individual success, and little or no emphasis on sophisticated technology and the acquisition of material wealth (Sorenson, 2003: 79).

African accomplishments before the European conquest

As stated earlier, ancient Africans accomplished a lot in the areas of philosophy, science, mathematics, engineering, commerce, trade, metallurgy, transportation, craft, education, agriculture, public and local government administration, architecture, and road building, to name but just a few, before the arrival the first Europeans on the continent in 15th century.

Philosophy

I recognize the folly of compartmentalizing the indigenous population of any locale by race and consider any effort to dignify the absurdity of such an exercise unnecessary. Here, I intend to do no more than simply state for the record that the original inhabitants of the general region of present-day Egypt were Kemites or Kemets. In fact, the initial name of Egypt was Kemet, after its original settlers. The reference in the above quotation to the 'Mediterranean white race,' was intended to convey the false notion that Egypt was and has always been settled by members of the white race. Those bent on deriding the contributions of the Black race to civilization deliberately ignore the preponderance of evidence attesting to the fact that the original inhabitants of Egypt, the Kemites, were members of the 'Black race.' The Kemites who have been consistently characterized in archaeological records as possessing typical Negroid features such as broad noses, woolly hair, and dark skin, migrated from an area closer to the equator to settle in an area along the southern portion of the Nile Valley. At its pinnacle, the Kemet Empire occupied an area a lot larger than present-day Egypt. Another group of Negro origin that counted among the original inhabitants of Egypt were the Nubians.

The Kemites and Nubians were philosophically profound and had established a well-developed system of thought and beliefs, complete with creeds and principles. They were amongst the world's first major civilizations. Some of their unique characteristics include the care with which they treated their dead. The dead were buried with great care in coffins lowered into lavishly equipped graves along with their personal treasures such as jewels. The bodies were buried to face westward, toward the sunset because they believed in life after death. A belief in 'Life after death' is one commonly held by Christians and other major religions worldwide.

Kemites lived by the virtues of *Maat*. These virtues comprise seven principles, namely Truth, Justice, Harmony, Balance, Order, Reciprocity and Propriety. To become one with Maat, the Kemites had to master the following (Michie, Online):

- Control of thought;
- Control of actions;
- Devotion of purpose;
- Have faith in the ability of [your] [teacher] to teach [you] the truth;
- Have faith in [yourself] to assimilate the truth;
- Have faith in [themselves] to wield the truth;
- Be free from resentment under the experience of persecution;

- Be free from resentment under the experience of wrong;
- Cultivate the ability to distinguish between right and wrong; and
- Cultivate the ability to distinguish between the real and the unreal.

The Kemites also operated a system of high moral and ethical values. Living according to the values means adhering to the forty-two-point Declaration shown in Table 2.1.

Table 2.1   The Kemet Declaration

01. I have not committed sin.	22. I have not polluted myself.
02. I have not committed rubbery with violence.	23. I have not terrorized anyone.
03. I have not stolen.	24. I have not disobeyed the law.
04. I have not slain men or women.	25. I have not been excessively angry.
05. I have not stolen food.	26. I have not cursed God.
06. I have not swindled offerings.	27. I have not behaved with violence.
07. I have not stolen from God.	28. I have not caused disruption of peace.
08. I have not told lies.	29. I have not acted hastily or without thought.
09. I have not carried away food.	30. I have not overstepped my boundaries of concern.
10. I have not cursed.	31. I have not exaggerated my word when speaking.
11. I have not closed my ears to truth.	32. I have not worked evil.
12. I have not committed adultery.	33. I have not used evil thought.
13. I have not made anyone cry.	34. I have not polluted water.
14. I have not felt sorrow without reason.	35. I have not spoken angrily or arrogantly.
15. I have not assaulted anyone.	36. I have not cursed anyone in thought, word or deed.
16. I am not deceitful.	37. I have not placed myself on a pedestral.
17. I have not stolen anyone's land.	38. I have not stolen that which belongs to God.
20. I have not been angry without reason.	41. I have not acted with insolence.
21. I have not seduced anyone's wife.	42. I have not destroyed property belonging to God.

Science, mathematics and engineering

The Kemites contributed enormously to the development of engineering, pure and applied mathematics, chemistry, geometry, medicine and pharmacology. They developed treatments for major and minor diseases and disorders as well as performed surgeries. Furthermore, they diagnosed, treated and indexed over two hundred diseases (Michie, Online). In the area of engineering, the Kemites also made tremendous strides by showing the world how to layout cities, build dams, canals, pools, ponds and gardens. They constructed great pyramids whose architectural complexity remains a marvel thousands of years following their construction.

The contribution of Egypt to civilization is legendary and hardly in dispute. What has however been at issue is, as shown above, the absurd question of Egypt's location. Egypt's breath-taking advancements in the areas of arts, science and technology have made it difficult for Europeans to concede that it is actually part of the vast continent of Africa. Conceding the absurdity of the question of Egypt's geographic location and unable to defend outlandish claims to the effect that Europeans were responsible for Egypt's contributions to civilization, Western historians have resorted to distinguishing between what they call 'Black' and 'White' Africans. Those involved in this desperate effort often end up attributing Egypt's success exclusively to the 'White race' and concluding, as noted earlier, that the 'Black race' has never made any contribution to civilization.

The Nubians, otherwise known as Ta-Seti or 'the Land of the Bow,' also made enormous contributions to the development of technology. Located south of Kemet, the Nubians were well known for the greatness of their citizens as archers and warriors. In the Old Testament of the Bible, Nubia is referred to on a number of occasions as Kush or Cush. The Nubians or Kushite Kingdom played a crucial role in the development of Egypt. In 1000 BCE Egypt's new Kingdom collapsed and Kush took over and became the dominant force in the region. The most prominent city in the nation of Kush was Meroe.

Meroe became economically powerful and dominated the entire region, with its sphere of influence extending to areas as far away as the Middle-East. Meroe was founded about 560 BC and located on the east bank of River Nile, some 200 km (120 miles) north of Khartoum, capital of present-day Sudan. It served as the capital of the Black Kingdom of Kush. Meroe, a great ancient African city was vaulted into prominence by its relatively sophisticated knowledge in iron technology. Iron technology, particularly knowledge of iron smelting, which had already reached an advanced stage in Africa prior to the colonial era, endowed some African cities such as Meroe with weapons and tools, thereby rendering them superior to their neighbours. Thus, it was not uncommon for communities with knowledge of iron technology to take over and/or control nearby villages and regions. Particularly because of its sophistication in iron technology, Meroe was able to control a number of neighbouring towns, including Axum and Adulis. It traded with these towns, as well as with Egypt and the Mediterranean world.

Arts, crafts and metallurgy

The dyeing, sewing, weaving, carving, sculpture, metallurgy and other artisan enterprises for which pre-colonial Africa is well known took place in towns and cities all over the continent (Hull, 1976). These cities were active, as opposed to passive, participants in regional economic, technological and cultural development process before the European colonial era. Particularly noteworthy in this regard is the fact that the socio-economic impact of African cities was felt in regions that were geographically far-removed from the continent. For instance, it was/is not uncommon to find bronze and terra cotta sculpture produced in the historical city of Ife, Nigeria, in far-away global cities such as New York, London, Paris, Tokyo, and Rome. The sculptors of ancient Ife employed a complex bronze-casting technique known as 'lost wax.' Incidentally, ancient Greek sculptors used this same technique. Perhaps this was sheer coincidence, or as some historians (e.g., Flint, 1966: 58) have suggested, the technique originated in Ife and found its way to Greece in the 12th century. How the artisans of Ife acquired the 'lost wax' technique in the first place remains enmeshed in mystery. To be sure, as Flint has argued, it is highly unlikely that Ife sculptors imported the technique from ancient Greece. A theory holding that the technique originated in Africa and found its way across the Mediterranean to Europe is not far-fetched in light of the fact that as far back as the 8th century A.D., Africa's exports to India and China already included iron (Mazrui, 1998; Ki-Zerbo, 1978).

Pre-colonial Africa's capacity for adaptation is particularly noteworthy. This capacity, Duignan and Gann (1975) have opined, enabled Africans to colonize a continent known for its harsh natural environment and climate. Pre-colonial Africans had succeeded, despite the limitations of the time, to develop viable techniques for working with the difficult soils and climates characteristic of the continent. Survival of the great ancient empires of Africa was also dependent upon their ability to defend themselves. Here, knowledge of metal or ironworks was critical for the production of weapons. Thus, as noted above, cities such as Meroe that were highly developed in this respect invariably constituted regional powers. The growth of city-states such as Ife, Benin, Old Oyo (in present-day, Nigeria), empires such as Ashanti (present-day, Ghana), Kongo (in present-day, Congo DRC and PR Congo) was due in large part to their industrial supremacy, particularly in the area of iron works, bronze and woodcarvings. Although seldom discussed, Africans before the European conquest had to have been skilled in mining otherwise, it would not have been possible for them to mine the gold for which the region was (is) known.

Trade and commerce

We hasten to note that large-scale socio-economic interactions have not always had a negative impact on the region. The negative impacts of such interactions were first felt in Africa in the 16th century, during the heydays of the transatlantic slave trade. We discuss the trans-Atlantic slave trade and other activities with negative implications for Africa below. For now, suffice to mention that the advent of the trans-Atlantic slave trade is perceived by many (e.g., Schneider, 2003) as marking the genesis of modern globalization in Africa.

Contrary to popular belief, the existence of towns or cities pre-dated the colonial era in Africa (Coquery-Vidrovitch, 1991; Anderson and Rathbone, 2000; King, 1990; Ki-Zerbo, 1978; Hull, 1976). Towns and cities constituted a crucial part of politically sophisticated and well-organized pre-colonial African kingdoms such as those of the Western Sudan, Ghana, Mali and Songhay. The political superiority of these empires and kingdoms was based on the technological advantage they commanded over other unorganized polities. For instance, the Great Empire of Ghana, which occupied a region to the northwest of present-day Republic of Ghana, established about 300 A.D., was able to conquer its neighbours thanks to its relative superiority in iron technology. Success in maintaining these kingdoms and empires depended on economically productive activities such as industry and trade. Trade in forest products accounted for the growth of the Songhay Empire, whose size at one point was the same as that of the continental USA and five times, that of the Holy Roman Empire (Mazrui, 1998). Geographically, the empire stretched from present-day Rio de Oro to Chad, and southward as far as to the forest line. It extended as far north as to present-day Algeria and Tunisia, and covered the entire south-western portion of Libya and almost reached the Mediterranean Sea near Benghazi.

A stronger determinant of the growth and survival of pre-colonial African towns as implied above is trade. Intense trading took place domestically, taking advantage of established regional and local networks. However, it was usually not enough for towns to be in contact with their neighbours. As Catherine Coquery-Vidrovitch (1991) notes, the international long-distance trade (across the Sahara Desert or Indian Ocean) proved invaluable to Africa's incorporation into the global economy. Those towns that showed an unparalleled degree of resilience were invariably involved, in one form or another, in intensive interregional and international commerce. The Kanem-Bornu Empire, which was located in the north-eastern portion of present-day Nigeria to the south-western tip of Chad and the south-eastern frontier of Niger, is said to have survived for as long as it did because of its strategic location at the southern end of the most easterly of trade routes from North to West Africa originating in Tripoli. Goods from the Mediterranean and the Arab world were dispatched through Murzuk and Bilma to Kanem-Bornu. Kanem-Bornu was effectively the point at which the trans-Saharan caravans terminated their journey, selling their goods of beads, glassware, salt, swords, cloths and lightweight manufactured goods. In exchange, the caravans collected valuable products such as hide, ivory, gold dust, copper ingots, iron, ostrich feathers, ebony, and kola nuts. Richard Hull (1976) identifies a number of cities, which later developed in the general area of Kanem-Bornu. These were a series of Hausa cities enclosed within clay walls. One of these cities, Katsina, which exist to date, boasted a population of almost 100,000 in the 18th century (Hull, 1976: xv). Towns in the Upper Zambezi region are also known to have participated vigorously in international commerce. They sold commodities such as wild rubber, ivory, gold dust, palm oil, wax, rhino horns, and in exchange, imported products such as muskets, cloths as well as luxury goods and tools through travelling traders from other regions.

A number of the prominent towns of sub-Saharan Africa, including Djenné, Timbuktu and Kano depended on crafts and commerce for their survival (Duignan and Gann, 1975). Thus, contrary to popular belief, pre-colonial Africa was not comprised exclusively of rural subsistence or peasant farming communities. The Hausa's located in the northern part of present-day Nigeria, for instance, imported virtually all their food by water-carriage along River Niger. In return, the Hausa's and the Sudanic communities of West Africa, who had made their mark as leather-workers and smiths, exported leather products, gold and clay pots. According to Duignan and Gann (1975: 3), towns in the region had attained a considerable level of economic specialization and 'their way of life had something in common with that of urban centres in early mediaeval Europe.'

To the southwest of present-day Nigeria lies the great historic city of Old Oyo. This city rose to prominence during the 16th century when it became an active participant in the trans-Atlantic slave trade, supplying thousands of slaves annually to Dutch slave buyers at Ouidah on the Atlantic coast in present-day Republic of Benin. In the same general area as Old Oyo, lies the historic city of Benin. Benin, unlike Old Oyo, was a commercial city-state, which dealt largely in forest products. Benin was not only an important commercial city. Rather, this great historic city and nearby Ife, were, as noted earlier, known the world over for the gorgeous brass and terracotta sculptures that their craftsmen produced.

African cities had been involved in such distant commercial relationships long before the European conquest. Duignan and Gann (1975) noted that pre-colonial Africa included societies, which possessed what at the time was definitely advanced knowledge in maritime travel. Some of the best known of the maritime cities, including Kilwa and Zanzibar in present-day Tanzania, Mombassa in present-day Kenya, Sofalla and Mogadishu in present-day Somalia, of that time were located on the eastern portion of the continent along the coast of the Indian Ocean. It is from these cities that Africans took off with local products in watercrafts bound for places as far away as the Arab world, India, and China.

Another important but oft-ignored factor that contributed to the growth of pre-colonial African towns was the ability of Africans of that era to adapt to changing times. Most notable in this regard was the ability to augment the limited stock of local food crops with imported varieties. Africans imported crops such as maize, cassava, sweet potatoes and yams from far away locations (Duignan and Gann, 1975). The adoption of literacy and Arabic script, albeit in a very small area and by very few people, rendered administration and governance more efficient (Flint, 1966; Duignan and Gann, 1975; Ki-Zerbo, 1975). Additionally, it facilitated the expansion of commercial and diplomatic contacts. In the more advanced urban cultures of Western Sudan and of East Africa, sophisticated credit systems predated the arrival of Europeans. In addition, trade currencies such as cowries and iron bars were widely in use.

Agriculture

From the beginning of times, Africans had learnt and perfected different methods of domesticating and cultivating food crops. The methods developed and handed down through generations have relied primarily on biological, renewable, local resources and knowledge. The methods, which have demonstrated their resilience by surviving through centuries, are ecologically tolerant, well-adapted to local conditions and include a diversity of crops. The system has guaranteed food security by dispersing plantings, heterogeneous genetic resources, minimum tillage and varying fallow. Ancient Africans perfected methods that took advantage of natural conditions such as the seasons of the year. In doing so, traditional methods of farming prove very conscious not only of the need to work in harmony with the environment, but also of the need for land to replenish itself. Thus for instance, traditional methods use no artificial manure and as a rule, allow fields to fallow after one or more crops. Mixed crop cultivation was also an important feature of traditional African agriculture. Success in agriculture hinged tightly on the traditional African land tenure system in which land is communally, as opposed to individually owned.

Success in the area of agriculture made it possible for pre-colonial Africans to indulge not only in local and regional trade, but also in international commerce. In this latter regard, pre-colonial Africans contributed to regional and international markets agricultural products such as wild rubber and oil palm. Perhaps more noteworthy here is the fact that participation in international trade was not limited to port and coastal towns. Rather, as Duignan and Gann (1975) noted, members of hinterland communities in pre-colonial Africa participated actively in long distance trade. To lend credence to their assertion, Duignan and Gann draw on the cases of Boer trekkers, who relied on travelling merchants for merchandise such as muskets and cloth. Similarly chieftains of Douala in present-day Cameroon, depended on palm products from the hinterland to play an active role in the regional and international palm oil trade.

Contact with far away regions, as implied above, played a critical role in the growth and survival of pre-colonial African towns. The growth and proliferation of towns such as Timbuctu, Djenné, Katsina, Kano and Bornu could have been unlikely without contacts with distant areas such as Europe to the north, the Arab world to the northeast and India and China to the east. Some historical accounts hold that the growth of towns and city-states in especially the Sahelien region is due at least in part to the importation of horses and camels, the wealth of the caravan trades, and Arab literacy (Flint, 1966; Duignan and Gann, 1975).

Transportation

By the time the first European explorers arrived Africa in the 1400s, Africans had domesticated some animals to help address the growing need of moving people, goods and services over land. At the same time, a number of innovations, such as the construction of rafts and canoes capable of providing water-based transportation services, had been made. Thus, the transport infrastructure in Africa at the time consisted largely, but not exclusively, of tracks for pedestrian and animal traffic, and natural navigable waterways. Some evidence suggests that a number of the ancient empires and city-states of the region had done well to develop a system of well-aligned roads and streets, as opposed to meandering footpaths. The ancient Ashanti Empire is said to have constructed an extensive network of roads that converged on the capital, Kumasi (Herbst, 2000). Griffiths (1995: 182) provided further evidence of the existence of a relatively extensive network of roads and water-based transport systems pre-dating the arrival of Europeans in Africa in the following statement.

> To reach Timbuctoo Gordon Laing followed well-established Caravan Routes across the desert from Tripoli. René Caille obtained passage on a boat, one of many sailing regularly up and down navigable inland Niger to Timbuctoo. Caille returned to Europe via the very old Caravan route through Morocco.

This extensive network of roads, footpaths and waterways later served to facilitate the transportation of slaves during the infamous transatlantic slave trade. The emphasis on road construction—albeit on a small scale and using very rudimentary tools commensurate with that era—that took place before any significant number of Europeans had arrived Africa, was abruptly terminated during the colonial era.

The colonial authorities were interested in penetrating the hinterland primarily to extract and transport raw materials to the seaports for onward transmission to the colonial master nations. Rail transportation presented itself as the optimal means of accomplishing this objective as well as that of militarily defending the colonial territory. For one thing, the cost of developing rail transportation facilities was far less than that associated with developing roads and concomitant facilities. For another thing, it was easier and cheaper to freight heavy and/or bulky goods by rail than by road.

A number of studies have unveiled convincing evidence attesting to the grandeur of Africa's towns pre-dating Arabic and European intrusions (Winters, 1983; Hull, 1976; Davidson, 1970). Although some (e.g., Coquery-Vidrovitch, 1991) criticize studies of this ilk, particularly Davidson's (1976), for exaggerating the so-called 'grandeur' of vanished African cities, there is mounting evidence suggesting that the cities in question had made technological, social and commercial strides deemed gigantic by the standards of that time (Winters, 1983). Thus, as Coquery-Vidrovitch (1991: 21) herself is quick to admit, '. . . precapitalist African urbanization did not need to be exaggerated or minimized, because its relatively restricted development fitted well with the demographic conditions and modes of production and long distance economic relations of mostly rural and trading societies.'

Urbanization and urban planning

Religion, particularly Islam, played a fundamental role in the growth and development of pre-colonial African towns. The Sudanic cities, such as Timbuctu, Gao, and Djenné, of pre-colonial West Africa are known to have been a popular destination of not only international traders but also Islamic clergymen from Arabia. In addition, these cities were not simply a popular destination for consumers of received culture. Furthermore, they did a lot to synthesize and transmit a wide array of cultures. The consumption of African spices, and use of ivory, leather, and other tropical products in the Arab world during that era lend credence to this assertion. The Swahili city-states of the coast of East Africa also attracted many visitors, such as Islamic clergymen and others from Arabia, Persia, India, Indonesia and the East African hinterland regions.

Long before Timbuctu, Djenné, and Gao rose to prominence, there was the stone-built town of Kumbi Saleh, which served as the capital of the ancient Ghanaian empire. Arab visitors to the region in the 11th century A.D. estimated that Kumbi Saleh contained more than 15,000 people and noted its vibrancy as a center of commerce. Kumbi Saleh soon became well known amongst Arab traders, who were interested in its spice, ivory and a host of other tropical products.

The regional and global reach of these cities at that time cannot be overstated. As the home of important Islamic institutions and advanced learning, Timbuctu had a significant influence on not only its surrounding regions but also on other areas of especially the Islamic world. The dominant architectural forms of Sudanic towns—clay structures with dome-shaped roofs—constituted the prototype for a significant number of human settlements in most Muslim sub-Saharan Africa.

No discussion of the historical importance of Timbuctu can be deemed complete without mention of Mansa Musa (nicknamed the Black Moses), who was King of the Empire of Mali from 1312–1337. As king, Musa catapulted Timbuctu to global prominence as a great centre of learning and Islamic scholarship. Timbuctu's prominence derived from the fact that it served as the venue for great scholars, poets and artists from Africa and the Middle East. Furthermore, it was home to vast libraries, *madrasas* (Islamic universities) and magnificent mosques. The largest of these mosques was built in 1320. Its architecture was impressive, comprising walls that reposed on plaster-like buttresses capped with finials, horns that projected from the walls to serve both aesthetic purposes and as scaffolding in the event of re-plastering or any other necessary maintenance. It is important to note that Islam, as it was practiced in the Kingdom of Mali at the time, had undergone serious adaptation with African religion, traditional beliefs, values and sensibilities helping to mould a uniquely African version of Islam.

King Mansa Musa was also well known for maintaining a huge military and police force, whose main duties included keeping peace and order along the trade routes; and maintaining the gold and salt mines. Mansa Musa is perhaps best known for the extravagant pilgrimage that he made to Mecca in 1325. Musa is said to have left Mali for Mecca with an entourage comprising 100 camel-loads of gold, each weighing 300lbs; 500 slaves, each carrying a 4lb gold staff; thousands of his subjects; as well as his senior wife, with her 500 attendants (Purpleplannetmedia, On-Line). To appreciate Mansa Musa's acumen in governance, one needs to only imagine the logistical nightmare inherent in managing more than 60,000 persons on a nine thousand-mile journey. Musa and his entourage made extravagant gifts of salt and gold to people in Egypt, to a point where the value of gold fell and remained low in Egypt for at least twenty years. It was thanks to King Mansa Musa's efforts that Mali appeared on the World Map in 1339 for the first time.

A number of towns, whose activities had implications for nearby and far away regions, were later to develop in the region encompassing the southern portion of present-day Burkina Faso, and the northern parts of present-day Côte d'Ivoire and Ghana. These towns included Bono-Mansu, Bobo-Dioulasso, Begho, and Kong. Amongst the inhabitants of these towns were Muslim traders who had established vibrant commercial and intellectual centers. Other known sojourners of the towns were Mande Dyula traders from Mali. These traders are particularly noteworthy because they were wont to travel by caravan southward in search of gold, which they sold to European explorers on the Atlantic coast. Recall that European explorers were increasingly arriving this region in the 16th century.

Africans before the European conquest also exhibited great skills and ingenuity in the area of urban design. For instance, one of the entries in a Dutch explorer's diary in 1602 made the following notation with respect to the ancient City of Benin (Njoh, 1999: 45, citing Tordoff, 1984).

> The town seemeth to be great; when you enter into it you go into the great broad street, not paved, which seems to be seven or eight times broader than Warmoes Street in Amsterdam; which goeth right out and never crooks . . .; it is thought that that street is a mile long [about four English miles]. . . . When you are in the great street aforesaid, you see many great streets on the sides thereof, which also go right forth.

Apart from their design prowess, ancient Africans were also cognizant of the importance of location as a determinant of the socio-economic development of human settlements. It is essentially the need to strategically locate its empire that prompted the leadership of ancient Ghana to move the empire from its original location in the western part of Sudan to its new site in the tropical rainforest and in close proximity to the region's two main rivers, the Senegal and the Niger. These rivers were later to play a crucial role in the empire's economic development and growth. Furthermore, the location placed Ghana at a critical trading node, thus permitting Ghana to control or serve as an intermediary in the trade in slaves, ivory, gold, and other tropical products from the south and salt, dried fruit, and kola nuts from the north.

Governance and administration

To survive, the huge empires of ancient Africa required a very functional system of revenue generation and financial management, a sophisticated administrative and governance structure and great leaders. Leaders were responsible for guaranteeing stability within the land. The African king was not, as he is often falsely portrayed in the West, a despotic ruler. Also, he could not act as an autocrat because he ruled with a council of elders. The king could be removed if he did not rule in the best interest of the people. The resemblance between this system and the U.S. system of governance that provides for checks and balance is striking. Furthermore, like the U.S. system, the African kingdoms provide(d) for the possibility of impeaching the king. Evidence also suggests that traditional Africans had developed complex and highly functional fiscal systems prior to the European conquest. For instance, the Ghanaian empire had developed a sophisticated taxation system, which charged duties on all goods that passed through the territory. Sandra Turner (on-line) has described this system in the following words:

> The king extracts a tax of one dinar of gold on each donkey-load of salt that enters his country and two dinars of gold on each load of salt that goes out . . . copper carries a duty of five mitqals and a load of merchandize, ten mitqals . . . All pieces of native gold found in the mines of the empire belong to the Sovereign, although he lets the public have the gold dust that everybody knows about. Without this precaution, gold will become so abundant as practically to lose its value.

It is important to note that salt in those days was equivalent in value to gold today— it was used as a preservative, seasoning and to prevent dehydration. The inland people did not have access to the salt mines. The sophisticated nature of the ancient Ghanaian governance is therefore obvious and incontestable.

One of the major achievements of ancient Ghanaians was their organization of the empire. The king ruled a functional hierarchy. He had a group of executive officers, his viziers; he had a governor to manage his capital city; and had as part of his court, the sons and heirs apparent of the kings and princes who were loyal to him. No doubt these princesses were present not only to enjoy the Court life but also as hostages to insure that their fathers remained loyal to the kings of kings. The African kings also had a crucial religious position. They were the intermediary between the people and their ancestors.

The Great Malian King, Mansa Musa also had a very efficient tax and financial management system that permitted him to maintain a vast standing army. He appointed to positions of governor, ex-generals with war experience. More than five centuries after Mansa Musa, the French colonial authorities were to employ a facsimile of this system in which they appointed military generals to the posts of governors in the colonies. The effectiveness and efficiency of King Musa's administrative and financial management system was put to a rigorous test in 1334 when it had to outfit thousands of people and animals for the king's historic pilgrimage to Mecca. The journey was more than nine thousand miles and lasted for more than one year implying a complex logistical task. Upon his return from Medina Musa successfully transformed Mali in general and Timbuktu in particular, into an Islamic centre of learning, attracting scholars from all over the Middle East and beyond. The most conspicuous of Musa's legacies is the monumental mosque, with seating room for 2,000 that he constructed in Timbuktu.

Another important historic figure, who presided over a large and complex kingdom and the large historic city of Djenné or Jenné, was Sunni Ali Ber. Djenné was a great city, which also boasted a university. The university was particularly known for its expertise in medicine and medical sciences. It also excelled in the area of surgery and medical research. The university was further known for its excellent architecture programme. Sunni Ali Ber, who died in 1492 has been likened to Napoleon Bonaparte for his military proclivities. He was a military genius who commanded and prosecuted his military missions with mathematical precision. He kept four palaces in operation simultaneously, governed with unparalleled efficiency, and appointed governors to oversee his provinces. He also recognized the strategic importance of the River Niger, as well as the need for fast mobile striking forces under the command of professional and well-disciplined soldiers. In short, like Bonaparte, Sunni Ali Ber established a semi-invincible military force.

Architecture and civil engineering

Elsewhere, I have observed that Europeans were wont to denigrate African architecture (Njoh, 1999). For instance, as recently as 1953, E.A. Gutkind, a noted urban historian had remarked that whatever Africans may consider their own architecture was devoid of 'a feeling of space as we understand it' and that 'Africans have never made an attempt to use space itself as a building material' (quoted in Njoh, 1999: 30). Gutkind's disparaging, and if I may add, disingenuous, characterization of African architecture is by no means unique. Rather, such characterization typifies interpretations of African scientific and technological innovations by cultures of prejudice. In some cases, cultures of prejudice have gone so far as to assign credit for purely African inventions or innovations to non-Africans. One example, that of the Great Zimbabwe Ruins (GZR) comes to mind here (Zimbabwe, On-Line). The GZR comprises a castle of interlocking walls and granite boulders that were constructed between the 13th and 14th centuries by wealthy Shona-speaking cattlemen in present-day Zimbabwe. It is believed that at its height, the GZR were full-fledged buildings that housed at least 40, 000 people. Some of the intriguing engineering features of the GZR include the dry-stone (i.e., no binding material used) walls, an array of chevron, herringbone and many other intricate patterns, and durability (the complex has survived almost completely intact for more than seven centuries). Due to the complex nature of the GZR, Europeans who are wont to trivialize African accomplishments have tried to no avail to theorize that Phoenicians, Arabs, Romans or Hebrews constructed the ruins. If such a theory has failed to gain any following, it is because there is no record of the presence of anyone from the named groups in the area prior to the European conquest.

Conclusion

One cannot but ponder why European explorers, slave traders and colonial and imperial authorities elected to ignore the enormous accomplishments of Africans, a minuscule portion of which I have identified and discussed above. It is no secret that Europeans consistently referred to Africa as the 'dark continent,' and to Africans as 'savages.' A lack of knowledge of Africa's accomplishments is certainly not one reason for this ostensibly hostile and racist valuation of Africa and Africans. Here,

I hasten to note that some of the major events in the history of Africa noted above coincided with a number of equally monumental events in the history of Europe. For instance, Sundiata's victory in the Kingdom of Mali (in 1235) came to pass within two decades of the signing of the Great Charter of English Victory or the Magna Carta, by King John on 15 June 1215. Also, as noted above, Mali had appeared on the World Map by 1339, a couple of decades before the first European explorers arrived Africa. Furthermore, it is important to note that the Great University of Sankore in Timbuktu was still an important centre of learning by the time the first European expeditions were initiated in Africa. Finally, it is worth noting that the death of Sunni Ali Ber, King of the Great Shonghay Empire in 1492 coincided with the so-called discovery of the America's or New World by Christopher Columbus. It was also about this same time that the first Portuguese explorers under the sponsorship of Prince Henry the Navigator were arriving the continent of Africa. It is important to note that Sunni Ali Ber's successor, Mohammed Touré otherwise known as Askia Mohammed served as vice chancellor of the University of Sankore and wrote at least 42 books in his lifetime. He is said to have restored the university's prestige by surrounding himself with learned people, especially from the Moslem world. In 1510 Leo Africanus characterized Mali's capital as 'a town of six thousand houses with several mosques and Koranic schools' and that 'the people of the town were merchants and artisans' with wit, civility and industry superior to all other Negroes (Turner, Online). To put things in perspective, we must note that European presence in the region date back to the late 1400s. The fact that the Portuguese built the Elmina Castle in Ghana in 1482 lends credence to my assertion. Thus, as I said earlier, Europeans had to have been aware of the fact that Africa was anything but a dark continent awaiting to be 'discovered' and 'opened up.'

Chapter 3

Colonialism, Christianity and the Erosion of African Culture

Introduction

The missions of colonialism and Christianity in Africa often overlapped and appeared almost always indistinguishable. Colonialism was instrumental in paving the path that enabled Christianity to penetrate Africa with amazing facility. As Richard Gray (1982: 60) noted, 'Christianity . . . made its rapid advances [in Africa] precisely because its emissaries, the missionaries, were so closely linked with the whole apparatus of colonial rule.' Although they never admitted this, Christian missionaries also had the responsibility of facilitating the achievement of colonial development objectives. In this respect, missionaries drew inspiration from carefully selected Biblical passages. Amongst the most prominent of these passages are the following from Jesus Christ's Sermon on the Mount: 'Blessed are the meek: for they shall inherit the earth' (Matt. 5:5); 'If someone strikes you on the right cheek, turn to him the other also' (Matt. 5:39); and '. . . love your enemies, bless them that curse you . . .' (Matt. 5:44).

The role of Christian religious authorities as facilitators in European efforts to exploit Africa did not begin during the colonial era. Rather, these efforts date back to the era of the infamous transatlantic slave trade. Slavery was legitimized and incorporated into the official *Corpus Iuris Canonici*, based on the *Decretum Gratiani*, which became the official law of the Roman Catholic Church under Pope Gregory IX in 1226. On 18th June 1452, Pope Nicholas V wrote the *Dum Diversas*, which granted the kings of Spain and Portugal the right to reduce all non-Christians to perpetual slavery. The issue of slavery was re-visited by the pontificate four decades later in 1493 when Pope Alexander VI issued the Papal Bull or the *'Inter Caetera,'* instructing Europeans to 'civilize' every 'savage' they encountered. The objective of 'civilizing,' I hasten to note, was also a central objective of colonialism. To rationalize or justify their perfidious schemes in Africa, the two institutions – that is, Christianity and colonialism – had to invent the myth of Africa as a dark continent and Africans as 'uncivilized,' 'primitive,' 'savages,' 'non-believers' (in God), 'animists,' 'idol worshippers,' 'fetish,' and 'pagans,' who had no knowledge of God. Working from this premise, European colonial authorities and Christian missionaries viewed their pre-ordained mission as essentially one of 'taming,' 'civilizing,' 'acculturating' and introducing God to Africans. In other words, the mission of these two institutions was to 'Europeanize' the African. Prosecuting this mission entailed activities with far-reaching implications for African culture and development.

My main objective in this chapter is to illuminate these implications. I do so by tackling the following three specific objectives. First, I expand on the discussion initiated in Chapter Two as a means of interrogating the contemptuous rationale for the metaphor of Africa as a 'dark continent,' peopled by 'savages' who had no knowledge of God. This metaphor, especially the fabricated and baseless theory that Africans possessed no knowledge of God prior to the European conquest, was necessary to justify efforts directed at destroying traditional African religious practices and beliefs. Second, I attempt to debunk this outlandish theory. Particularly, I marshal evidence demonstrating that the knowledge of God and the concomitant creation myth were prevalent throughout the continent. Interestingly, some versions of the African creation myth bear a striking resemblance to the dominant orthodox Christian version as narrated in the Bible. I spend a substantial part of the chapter shedding some light on traditional African religious beliefs and a few versions of the creation myth. My final objective in the chapter is to show how the activities of Christian missionaries and colonial authorities conspired to stifle development in Africa.

Traditional African beliefs and knowledge of God

The following two important questions have often constituted the subject of rancorous debate amongst students of culture and religion in Africa. Were Africans a religious people prior to the arrival of Europeans on the African continent? Did the knowledge of God precede the arrival of Europeans in Africa? European Christian missionaries and colonial authorities on the continent, in line with their metaphoric characterization of Africa as a 'dark continent' recorded that Africans had no knowledge of God and no religion worthy of any attention. To European explorers, colonial authorities and Christian missionaries, the traditional African was essentially a 'childlike' individual who lacked what it takes to wrestle with complex questions relating to the origins and nature of humans, the universe, life forms, death as well as other intricate puzzles such as the raison d'être of human existence, life or the lack thereof after death and so on. How representative of the 'true' traditional African was this picture? Is there such a thing as 'traditional African religion' that predated the introduction of Christianity, and Islam before it, on the continent of Africa? Were traditional Africans cognizant of the existence of God or an Almighty Being? To address these questions it is necessary to shed some light on the concept of religion.

The word 'religion' in English may be rooted in the Latin word *'religo,'* which can be roughly translated to mean *'ritual'* or another Latin word, *'religãre,'* meaning *'to tie fast.'* In general usage, religion is usually taken to encompass beliefs, feelings, doctrines and practices linking a people (usually a community of believers) to a sacred and higher level spiritual being. Definitions such as this have been criticized for their propensity to exclude establishments or systems of beliefs and practices that a significant number of people passionately view as religious but which do not recognize or believe in a personal deity or some supernatural or higher order being (e.g., non-theistic religions such as Budhism and religious Satanism) (Religious Tolerance.Org, On-line). One such critic, a group that goes under the name 'Religious Tolerance' defines religion broadly to include 'any specific system of belief about deity, often involving rituals, a code of ethics, a philosophy of life and a worldview' (Religious Tolerance.Org, On-Line). The notion of worldview in this regard refers to basic foundational beliefs about humans, the rest of the universe and how both are linked to a supernatural being.

On account of this definition, it is clear that Africans had established religions and were a profoundly religious people who possessed a sophisticated knowledge of God before the European conquest. A comparison of how traditional Africans viewed God vis-à-vis the concept of God in Christian thought lends credence to this assertion. Christian doctrine personifies God. According to Christian teachings, God created man in his own image. It follows therefore, that man bears some resemblance to God. In fact, God, according to Christians is a man. Initially, He created one man, called Adam (in His own image) and the idea of creating a woman, Eve, came only as an after-thought. Both Adam and Eve were, according to all Christian portrayals, Caucasians or 'white' people. In a further effort to personify God, Christians believe that He worked for six days (creating the universe and everything therein) and took the seventh day off. Also, they believe that He lives in a specific place called Heaven, which is literally above the earth and beyond the skies.

While some traditional African religions attribute the male gender to God or the Supreme Being, a few believe that God is female while most attribute no gender to God. For instance, as Ifeka-Moller (1974: 60) noted, amongst the Ibo's of Nigeria, the Ikwerri Ibo's of Ahoada Division believe that the Supreme Divinity, *Chi*, is a female, while the northern and western Ibo located on the east and west banks of River Niger believe that the Supreme Divinity, *Chi-uku*, is a male. The Meta of the southwestern edge of the Bamenda plateau in Cameroon exemplify traditional African groups that possessed a conception of a Supreme Being who was 'sexless' prior to the introduction of Christianity and other alien religions in Africa.

To be sure, the traditional African view of God differed and even contrasted sharply with the Western view in many other respects. For instance, Africans found efforts on the part of Christians to personify God as absurd and ridiculous at best. In traditional African religious thought God bore no resemblance to humans. This meant, *inter alia*, that He did not take the 'seventh day' off as Christian missionaries preached. God, in African religious thought is mysterious, has never and will never take time off. Africans believe the consequences for the world would be catastrophic if God were even to blink, let alone take time off. In fact, the notion of the seventh day made no sense to Africans, who observed weeks with a varying number of days. For instance, the Meta's and almost all groups of the North West Province of Cameroon observed an eight-day week while the ancient Egyptians observed a ten-day week prior to the European conquest. Most traditional Africans tailored the duration of a typical week based on market days. Thus, eight- or ten-day weeks were considered apropos as they allowed farmers sufficient time to harvest farm products or prepare and transport handicrafts or other products to be sold at local and regional markets. Ancient Egyptians avoided the seven-day week as the number seven was deemed to be associated with 'bad luck.'

The skepticism of Africans with respect to the tale of creation was well founded when one considers the fact that the seven-day week, which is part of the Gregorian calendar, has its roots in Christianity. Pope Gregory XIII, head of the Roman Catholic Church, which at the time was the official religion of the Roman Empire, decreed the Gregorian calendar on 24 February 1582. Another Roman leader, Emperor Julius Caesar, introduced the forerunner to the Gregorian calendar, the Julian calendar in 46 BC. Vincent Malett (On-Line) paints a more lucid picture of the Christian origins of the seven-day week when he contends as follows.

> The 7-day week was introduced in Rome (where ides, nones, and calends were the vogue) in the first century A.D. . . . when Christianity became the official religion of the Roman empire in the time of Constantine (c. 325 A.D.), the familiar Hebrew-Christian week of 7 days, beginning on Sunday, became conflated with the pagan week and took its place in the Julian calendar.

It is important to note that Africans were not the only ones who did not function on a seven-day week schedule prior to the widespread use of the Gregorian calendar. The Maya calendar, for instance, has a thirteen- and twenty-day week. Similarly, the Soviet Union (prior to 1929/30) had a five- and six-day week, which was replaced by a five-day week in 1930; and then by a six-day week up until 26 June 1940 when a seven-day week was decreed. Lithuanians operated on the basis of a nine-day week before Christianity was introduced in the country.

Another element of Christian religious thought that most Africans viewed as absurd relates to the belief that God had or has helpers (angels). Within the context of African religious thought, God is so powerful and mysterious that He has never needed, and does not need, anyone to lend Him a helping hand. Accordingly, He single-handedly created everything in the universe and beyond. The Gikuyu of Kenya have a concept of God that mirrors that of most Africans. The Gikuyu believe that God or *Ngai*, the Creator and giver of all things, has no father, mother or companion of any kind (Khapoya, 1998). However, it is necessary to acknowledge the fact that some, albeit few Africans, such as some Ibo's and Yoruba's of Nigeria believe that God is assisted in His daily functions by smaller deities. Yoruba's of this persuasion refer to the lesser deities or the Supreme Being's aides as *Orisha Nla*, while the Ibo's harbouring this notion call them *Orishas*.

African religion, God and the creation legend

Legends or myths meticulously describing the origins of the universe abound in Africa. A few interesting examples, some of which almost mirror the Judeo-Christian or Western creation tale, are in order. I focus particularly on the following six: the Yoruba of Nigeria, the Bushmen of Kalahari Desert, the Zulu of South Africa, the Bantu of the Central African region, South-Eastern groups of Nigeria, and the Shiluks of the Nile region (Metmuseum.org; Klitz, On-Line).

The tale of creation in Yorubaland

The Yoruba creation myth holds that the Supreme Deity or God of all gods *(Olurun)* created everything that is and ever shall be. He did so with the able assistance of lower gods. The chief of the lower gods is known as *Orisha* or *Orisha Nla*. *Olurun* assigned *Orisha Nla* the task of molding human figures out of clay. However, only He, *Olurun*, gave humans life, which He did while all His helpers, including *Orisha* were in a slumber. The Yoruba's believe that the center of the universe is Ile Ife or 'Wide House.' It is from here, according to the Yoruba legend that all human beings originated and then spread out to other regions.

The Bushmen tale of creation

The Bushmen of the Kalahari Desert believe that, the universe and everything therein, beneath and above it, was created by Kaang, the Great Master and Lord of All Life. According to the legend, Kaang initially placed all living things, including animals and people underground. These creatures lived in harmony, did not have to work, and communicated with each other in a common language. Then, one day, because the humans broke one of Kaang's rules, he decided to bring them above ground. He began by creating a large tree, then he drilled a hole to permit the creatures access to the surface. He then began helping the man and the woman, and then the various animals and other creatures to the surface one by one. Once above ground, the various creatures began developing different unique ways of communicating. Also, they were required to fend for themselves.

Bushmen believe that Kang gave the same life to humans that he gave everything else, including rain, thunder, wind, trees, animals and so on. This life is in the form of spirits, which cannot be seen. What is visible to the naked eye is just a shell or body. Within these shells, including those of humans are spirits. These spirits can leave from one body form or type to another. For instance, the Bushmen believe that a human's spirit may take up residence in a lion's body and vice versa.

The Zulu creation myth

The Zulu of Southern Africa recognize the existence of the Almighty Being, which they call the *Unkulunkulu*. The Zulu's believe that the creator originated in the reeds and it is from there that he brought forth the people and cattle. According to the Zulu's, the creator, created everything in the universe. Furthermore, he taught them not only how to hunt but also how to make fire and grow food.

The Bantu creation story

The Bantu's of the central African region believe that in the beginning there was only the Supreme Being, *Bumba* and a world full of water but without light. Then, one day, *Bumba* came down with a case of stomachache, which caused him to vomit. The vomit turned into the sun, which dried up some of the water leaving in its wake, land. *Bumba's* stomachache continued, causing him to vomit again. This time, the vomit transformed into the moon, the stars, and then some animals, and then people.

The story of creation in South-Eastern Nigeria

Some major groups in this region refer to the creator as *Abassi*. *Abassi* created two humans, a man and a woman, and then decided against them inhabiting the earth. However, his wife persuaded him to reverse his decision, which he obliged on condition that the two humans will eat all their meals with him. This was meant to prevent the humans from hunting or cultivating food. The humans were also forbidden from procreating. However, against *Abassi's* orders, the woman began growing food and soon, the two humans stopped showing up for meals with *Abassi*. Later on, the man joined the woman in the field and soon thereafter, they began having children, again, in contravention of the orders of *Abassi*. The creator, *Abassi* blamed his wife contending that were it not for her the humans would not have inhabited the earth in the first place. *Abassi's* wife reacted by sending death and discord to earth as a means of punishing people for disobeying *Abassi*.

The creation story of the Shilluks of the Nile region

The Shilluks believe that human beings were molded from clay. The creator is believed to have been a very extensive traveler, who visited every part of the world, creating humans from the clay and dirt in each region. The Shilluks believe that it is precisely for this reason that humans possess different skin colour – black, red, brown, white, and so on. It is also for this reason, they believe, that humans from different regions have different hair colour and texture. Perhaps more importantly, the Shilluks note, is the fact that people from different parts of the world have different cultures, norms and behaviour.

African belief systems

The notion of God possessing an abode located in a special place called Heaven went against the grains of African religious thought, which held that God was everywhere or omnipresent. Similarly, the Christian belief that God was a man conflicted harshly with the tenets of African traditional religious thought, which hold that God can assume the form of a spirit or any object, including a tree, a rock, a body of water or even a neighbour or total stranger. This assumption is at the root of traditional practices that condemn any actions with negative implications for other people and the natural environment. The Africans see God as so mysterious that He was capable of simultaneously taking on the mentioned and other forms. He can at once be the 'God of Fertility,' the 'God of Fortune,' the 'God of the Forest,' the 'God of Rain,' the 'God of Rivers or Water,' the 'God of the Sea,' the 'God of Fire' and so on. Traditional Africans often invoke the name of God in His various capacities to intervene in their lives as need be. For instance, in periods of droughts, traditional Africans are wont to invoke the name of God in His capacity as rainmaker – that is, the God of Rain or God of Water – to intervene. Similarly, when a traditional African family is faced with the problem of infertility, it prays for intervention by God in His capacity as God of Fertility. Traditional Africans are also wont to pray to God in His capacity as the God of fertility to make the soil productive during the planting season and for bountiful harvests during the harvesting season.

This conception of God in multiple capacities and even multiple 'persons' has parallels in Christian thought which early Christian missionaries conveniently ignored in their dealings with Africans. The Bible, talks of the God of Abraham, the God of Moses, or God in three persons, and so on. This notwithstanding, Christian missionaries derided and chided Africans for worshipping multiple gods. For instance, Khapoya (1998) remarked that the tendency on the part of the Kipsigis, a subgroup of the Kalenjin people of Kenya, to refer to God in His multiple capacities led Christian missionaries and Western scholars to erroneously conclude that the Kipsigis worshipped several gods and possessed no knowledge of God Almighty.

Traditional Africans believe in spirits. This tendency, the belief in spirits, is an important attribute of the traditional African religious system. Spirits are thought of as important but invisible forces that are capable of functioning on their own orders or those of human beings. Spirits in the context of African religious thought can be divided into two major groups, namely 'nature spirits' and 'human spirits' (Khapoya, 1998). Nature spirits are associated with 'real' objects – that is, objects that can be seen, felt and/or touched.

Examples of such objects include the sky, the moon, rain, sun, trees, rivers, mountains, and rocks. The spirits taking residence in any of these objects may be male, female or unisex. For instance, in most of West Africa, the most prominent spirit residing in rivers or other bodies of water is commonly known as *Mammy Water*. Spirits are believed to come out of their residence every once in a while to reprimand humans for stepping out of line or violating a natural or other order. For instance, mountains spewing out lava that causes irreversible damage to buildings, crops, and humans in the immediate vicinity during volcanic eruptions may be perceived as reprimanding humans for terrible deeds such as environmental abuse and degradation. Traditional Africans believe that failure to appease the river spirit, *Mammy Water*, may lead to damage or accidents on any bridge over the river. Similarly, it is believed that failure to appease the soil spirit of a certain region may result in poor crop yield or high accident rates on roads constructed in that region.

The second group of spirits is known as 'human spirits.' Human spirits originate in, or are associated with, the deceased. Usually, these are the spirits of relatives who have passed away either recently or as is usually the case, in the distant past – that is, some generations ago. In the context of the traditional belief system, every living thing or creature in the universe possesses a spirit. It is believed that a creature's spirit survives and lives on after the creature is dead. It is further believed that the spirits of the dead occasionally make their presence felt by the living. Such occasions may arise when the living commit a taboo or violate a societal norm. In this case, it is believed that the spirit of the dead may assert its presence and viability by punishing the living. Human spirits are believed to express their discontent with the behaviours of the living by causing terrible or unpleasant things to happen to them and/or members of their families. Once this occurs, the living are required to perform certain rituals.

However, spirits are not only known for executing projects with negative consequences. Rather, they are also believed to indulge in acts of benevolence. Accordingly, traditional Africans are also wont to invoke the spirit of their dead relatives and ancestors to help them resolve complex and difficult problems ranging from defeating a formidable enemy in a war or other form of altercation to rendering a family fertile and reproductive.

As stated earlier, once an infraction has occurred or a taboo has been committed and the spirits of the dead have sanctioned the living by punishing or reprimanding them in some manner, shape or form, a ritual of some form must be performed for things to return to normalcy. Rituals are also necessary to invoke the spirit of the dead to help resolve any quandary that the living may be facing. Such a quandary may have to do with a negative development in a family such as a member of the family taken ill or a baby developing rather too slowly or refusing to take breast milk for reasons that are not readily obvious.

Rituals in the religious thought of Meta, Cameroon

Two examples from Meta in the North West Province of Cameroon will do well to illustrate the two scenarios implied above. The North West Province of Cameroon, sometimes referred to as the Bamenda Grassfields has been extensively studied by anthropologists since the colonial era (see e.g., Chilver and Kaberry, 1966; Chilver and Kaberry, 1968; Chilver, 1963; 1966; 1967a; 1967b; 1989; Drummond-Hay, 1925). The country of the Meta (also referred to as the Menemo) is located on the southwestern edge of the Bamenda plateau. According to very early colonial census records (circa, 1900), the Meta had a population of about 20,000 (Dillon, 1973; Gregg, 1924). According to Richard Dillon (1977: 155), one of the few anthropologists who have conducted extensive research of the Meta, the social organization of traditional Meta society 'was founded on patrilineal descent and village organization.' Dillon draws attention to an aspect of traditional Meta society that is particularly of relevance for the purpose of the discussion in this chapter. This aspect has to do with the fact that traditional Meta society was comprised of the following.

Localized patrilineages [which] regulated the inheritance of farmland and the distribution of bridewealth, and performed important rituals designed to ward off illness and misfortune from their members. Most lineages were also linked by ties of clanship to comparable groups scattered throughout Meta territory (Dillon, 1977: 155).

The first case is one with which I am personally familiar. It involves a baby who had suddenly refused sucking at the mother's breast after having done so without a hitch for the first fortnight subsequent to his delivery. Two days after all efforts to get the baby to suck at the mother's breast proved abortive, a diviner was consulted. The diviner's diagnosis traced the cause of the baby's abnormal behaviour to the fact that a portion of the dowry on his mother had not been settled. The owed portion was a payment in the form of a goat that the baby's paternal uncles were supposed to have made to his maternal uncles. The diviner instructed the baby's paternal uncles to immediately make the necessary payment to prevent the baby from starving to death. Unfortunately, the said uncles had no money to acquire a goat, which is an expensive commodity in Meta country. However, and fortunately, traditional Meta traditional law is pretty lenient towards debtors. A debtor who is unable to meet his obligations to his creditor(s) may simply acknowledge his indebtedness with a token serving as a promissory note that the actual payments would be made on a date in the future specified and agreed upon by him an his creditor. Acceptable tokens vary with the debt in question. In the case of a goat, the acceptable token is usually a rope of the genre used for tethering goats. The rope must be knotted in a special fashion and presented to the creditor.

Thus, the baby's paternal uncles were instructed to seek and present such a rope to the baby's maternal uncles. On the evening of the third day of the baby's refusal to suck at the mother's breast, his paternal uncles took, as per the diviner's instruction, the rope and a calabash of palm wine and traveled several miles to his maternal uncles. At about the same time that the negotiations between the baby's paternal and maternal uncles were taking place, something miraculous happened – the baby resumed sucking at the mother's breast again. The paternal uncles returned home late that evening to be greeted by the good news.

The second case comes from Richard Dillon (1977) and it has to do with the ritual of curing and conflict resolution in Meta country. In the particular case in question, an old woman, the widow of a recently deceased member of a lineage was taken seriously ill. In addition, the woman's brother in-law, who, according to Meta custom, was the heir to the family estate, had developed a drinking problem and was becoming increasingly irresponsible. A diviner determined that the source of the woman's illness as well as her brother in-law's social problems was a curse (known in Meta language as *ndon*) from ancestral spirits. In the context of Meta religious thought, *ndon* is a type of supernaturally induced ill luck, which may manifests itself in the form of illness, poverty, infertility, and even death, *inter alia*. Dillon (1977: 158) further notes that *ndon* may be triggered by either a dramatic event, such as a homicide or fight within a lineage, or by the complaint of a person whose rights have been seriously offended. In the latter event, the likelihood of the offender suffering *ndon* is thought to increase when many people, in addition to the injured party, join in discussing the offence. This brings the matter more quickly and surely to the attention of *Nwie*, the Creator Spirit, who ultimately inflicts on man the sanction of *ndon*. *Ndon*, however, need not always strike the offender himself. Frequently it affects in his stead other members of the household or lineage.

In the case in question, the diviner traced the roots of the *ndon* to one specific source. The source was a long-standing dispute between two deceased members of the lineage, one, the former lineage head and the other, the late husband of the afflicted widow. The two deceased parties had a bitter and violent quarrel that involved threats of homicide – a taboo in Meta tradition – before passing on. According to Meta religious thought, the commission of such a taboo automatically invokes *ndon*, a situation that can only be reversed subsequent to the performance of certain specific rites. Typically the prescribed rituals take place at two levels. The first level involves slaughtering and sharing a sacrificial goat amongst designated non-family members of the clan, and preparing and serving these individuals food and drinks. The second phase involves repeating this process for family members. These rites are meant to efface the *ndon* and re-instate cordial relations between the parties and/or families involved. Until the rites have been performed, members of the quarrelling families are forbidden from interacting with each other.

In the case under consideration, only the first phase of these rites had been performed. Yet members of the two families, especially the young adults and children, were interacting with each other in contravention of Meta religious doctrine. The diviner pinpointed this as the source of the problem and recommended that the families immediately embark on the process of accomplishing the second phase of the rites. The families heeded the diviner's advice. Part of the ritual, which as stated earlier, involves preparing and serving the concerned parties food and drinks, had to do with acquiring and sacrificing a goat at the lineage's sacrificial site. The specific steps involved in this process are telling and deserve being described in some detail here. Dillon (1977: 162) described what took place in the particular case in question as follows. 'Arriving the sacrificial site, the men first removed the goat's tethering rope and untied the knots in it to signify the freeing of the lineage from *ndon*.'

Although the scope of Dillon's study did not permit him to hang around enough to determine the impact of the ritual on the widow's illness as well as her brother-in-law's problem of drinking and social irresponsibility, as someone who has firsthand knowledge of Meta tradition, I can state without equivocation that the results of such rituals are often mixed. Sometimes these and cognate rituals precede the desired state of affairs. Thus, the relationship between the rituals and the desired state of affairs do, at least in some instances, meet one criterion for inferring causality in the social sciences. In the social sciences, it is necessary, although not sufficient for cause to precede effect to infer causality.

Belief systems, implications for development

The colonial era in Africa marked a period in the continent's history when ancient cultures and traditions – viewed by colonial authorities and other Western change agents as primitive and antithetical to development – were brought face to face with so-called modern civilizations. The notion of African cultures and traditions as primitive constitutes an important element in the metaphor of Africa as a dark continent. As Lucy Jarosz (1992: 107) has observed and I concur, this metaphor paints a visceral and compelling image of Africa as 'untamed, unknown, and evil.' More importantly for the purpose of the present discussion is the implication that the continent was badly in need of enlightenment – a feat which colonial authorities and Western change agents considered themselves uniquely qualified to accomplish. Enlightenment was operationalized in several ways, two of which are relevant here. First, enlightenment was seen in terms of Christian missionary activities, particularly those with the immediate aim of converting Africans to Christians. In this case, the appellation 'Dark Continent' symbolized neither a place nor a person. Rather, it signified the absence of Christian belief and practice. Christianity was posed as the light designed to dispel the darkness characterized by African indigenous religious beliefs and practices. Within this framework, Africans were branded as superstitious pagans who worshipped mountains, trees, rivers, rocks and everything but 'Jehovah' or 'the true God.' In this regard, one commentator writing during the early days of the European colonial era in Africa acknowledged the fact that Africans were aware of the existence of a 'Great God above all gods,' but asserted that the 'Great God' in question 'is not Jehovah, nor a reminiscence of Jehovah' (Kingsley, 1897: 141). Thus, the darkness in the metaphoric 'Dark Continent' of Africa assumed a particular conceptual stance for missionary objectives and activity on the continent. From this stance Africa was 'dark' and in need of 'light;' it was 'tribal' and in need of 'Christian ideas;' it was 'lost,' and needed to be 'found' (Jarosz, 1992). Second, enlightenment was operationalized in terms of 'opening up' the dark recesses of the continent to scientific discovery as well as incorporating the entire continent into the global capitalist system. This was in line with the imperialist, civilizing and acculturation missions of the colonial project.

As Jarosz (1992: 108) noted, Africa was thus characterized as a 'primeval, bestial, reptilian or female entity to be tamed, enlightened, guided, opened, and pierced' by Europeans. What measures did Europeans adopt in their efforts to 'enlighten,' 'tame,' 'guide,' 'open,' and 'pierce' Africans? How did the resultant changes affect the continent's development prospects? I spend the remainder of this chapter attempting to address these questions.

The most prominent tool in efforts on the part of Europeans to supplant the indigenous religious practices and beliefs of Africa with European varieties, particularly those rooted in Christianity and Western ideals, was formal education. In this regard, colonial governments worked in collaboration with Christian religious bodies. Christian mission schools received financial assistance from colonial governments on condition that they adhered to school programmes that contributed to the attainment of colonial development objectives. Programmes were required to focus on the formation of a logical mind in African children, by destroying what they (colonial and Christian missionary authorities) viewed as superstitious beliefs, and establishing in their stead, rational and scientific relations between cause and effect. Christian mission schools included instructions in Christian religious doctrine. Formal educational programmes impressed upon Africans that Western religious thought is scientific while the African variety was not. African minds, it was believed tended to obey laws that had nothing in common with the rational ways professed to characterize European civilization. Westerners thus derided African religious beliefs and practices as superstitious and primitive.

Yet the religious beliefs and practices of Africans were not any further removed from rationality than those of Western Christians. By its very nature, religion, including Christianity, is grounded in faith – faith that is usually based on dogmatism requiring neither hard evidence nor verifiability. It is no secret that Christians believe that faith and prayers have the power to heal, reverse undesirable situations, bring fortune to 'believers,' transform wishes into reality, and so on. Christians also use prayers, which they typically do whilst invoking the names of lower level deities such as angels and saints, as well as ritualistically employing Christian symbols such as the crucifix, the statue of the Virgin Mary, rosaries and so on, as a means of communicating with Jehovah, the Almighty God. In an identical fashion, traditional Africans ritualistically use African religious symbols such as rocks, and animal horns as well as indulge in rituals such as the pouring of libation while invoking the names of their ancestors as they communicate with God the Almighty Creator.

From a positivist perspective, we are just as incapable of verifying the effectiveness of Christian rituals (e.g., prayers) as we are of verifying the effectiveness of African traditional religious rituals. Members of each of these two groups have endless tales to narrate about the success of their rituals. A Christian that wins a lottery, passes a very difficult test, successfully undergoes a surgical operation, or escapes a ghastly automobile or any other accident unscathed is likely to attribute her fortune to her close relationship with some saint, guardian angel, Jesus Christ and ultimately with the Almighty God. Similarly, an African involved in the same situation is likely to credit her ties with her ancestors and ultimately with God Almighty with the fortune. However, the scientific mind is likely to disagree as each of the foregoing scenarios cannot be scientifically explained. Are we therefore to conclude that rituals have no impact on outcomes? The response to this query cannot be in the affirmative in the case of Christianity or any other major religion for that matter but in the negative for African traditional religion.

Let us re-examine the two rituals culled from Meta society in Cameroon and recounted above. The outcome of the ritual in the case of the baby who had refused to continue breastfeeding at the tender age of two weeks was positive. So too was the ritual in the case of the ailing widow. How are we to explain these outcomes? To the families involved in the two cases, there is no doubt that the rituals were directly responsible for the positive outcomes. I tend to disagree. However, I submit that there is an indirect relationship between the rituals and the positive outcomes.

To appreciate this link in the breastfeeding case, it is important to acknowledge the fact that according to Meta traditional religious doctrine, children born to a woman whose in-laws have not met all their obligations to her family may fall sick and/or die. The mother of the baby in this case was fully aware not only of this doctrine but also of the fact that her in-laws had not met all their obligations to her family of birth. This knowledge was quite possibly a cause for consternation for the young mother. To the extent that a nursing mother's ability to produce breast milk hinges tightly on psychological factors, especially her state of mind, it is arguable that her constant concern plausibly rendered her unable to produce breast milk. It is conceivable that the baby stopped sucking after doing so was no longer yielding any dividends.

In the second case, that involving the ailing widow, Richard Dillon (1977) shares my opinion that the positive outcome can be attributed, at least indirectly, to the rituals that were performed. In this case, the rituals can be seen as a process designed to reassert traditional social order, promote social solidarity and reaffirm traditional norms and values. Here, it is important to note the following (Dillon, 1977: 166).

> First, during the rites the participants were gradually drawn together so that by the end, they showed relatively greater unity and agreement. Second, key social values, such as respect for the elders, the need for cooperation within the lineage, and the importance of responsible behavior by young male successors, were repeatedly stressed by participants in the ritual.

An important role of lineage heads in traditional African society is to cater to the well-being of family members. Thus, it is conceivable that in the case described above, the absence of solidarity in the family as well as the irresponsible behaviour of the young male successor could have created a stressful environment that contributed in no small way to the widow's ailment. It is equally conceivable that the ailing widow's condition could not truly be resolved until the male successors assumed their responsibilities as stipulated by Meta custom and tradition. To the extent that this came to pass, we can explain the widow's recovery as a function of the improved family relations (occasioned by the rituals). Thus, if nothing else, the speeches and discussions that constituted part of the rituals succeeded in reminding the male successors of the need to recognize and fulfill their obligations to the family. One of these obligations is providing the funds necessary for treating ailing members of the family in particular and the lineage in general.

Development implications

Lucy Jarosz (1992: 105) has persuasively argued that the metaphor of Africa as a Dark Continent, inhabited by uncivilized people who worshipped inanimate objects and had no knowledge of God was designed to 'reaffirm Western dominance and reveals hostile racist valuations of Africa and Africans.' More importantly, this metaphor was used to justify efforts that sought to transform the continent into a pseudo-facsimile of the West. These efforts contributed enormously to the accomplishment of European colonial and imperial objectives but stifled socio-economic development in Africa.

Prior to the advent of colonialism and Christianity in Africa, religion was a way of life throughout the continent. African traditional religion was not organized in the manner of Christianity. The African takes his religion with him everywhere he goes: to the farm, to social events, and other functions. The imposition of colonialism and Christianity was accompanied by many significant transformations, including a new concept of the week. A week was no longer eight or ten days long. Rather, it became seven days long with one day decreed as a day when nothing but Christian-style worshipping had to be done.

Colonial governments throughout Africa adopted the Gregorian calendar with the concomitant seven-day week format. Accordingly, businesses and offices were typically open all-day from Monday through Friday, and half a day on Saturday. Sunday, which Christians believe was the day God took off, was set aside for Christian-style worshipping. The imposition of this Western week format effectively obliterated the functioning of African traditional institutions. For instance, it made it difficult at best and impossible at worst for Africans to participate in traditional markets, especially because the days for such markets coincided mostly with official weekdays as opposed to weekends. Yet traditional markets provided a forum for socializing. Such socialization played an important role in the functioning of African society. The traditional African market square is a venue for traders, members of extended families, and friends to meet and interact. In traditional Africa, most issues and/or events of relevance to families, in particular and communities in general are planned at market squares.

Organized religion, particularly Christianity also meant that African converts had to set aside several days in the shorter – seven- as opposed to eight- or ten-day – week to participate in church events such as choir practice, and other monthly or annual events such as preparation for communion, thanksgiving, mission feasts, and so on. Added to this was (is) the direct financial cost of being a Christian. Such costs can be seen in terms of the annual dues Christians must pay just for being members of any given Christian congregation, the additional financial costs associated with membership in Church groups such as Christian Men's or Women's Fellowship, Catholic Youth Association, Christian Youth Fellowship, Choirs, and so on. The negative implications of these financial burdens and for dedicating several days in any given period to a single activity – in this case, religion – for agrarian societies cannot be overstated. I hasten to note that Christians in the West do not commit nearly as much in terms of amount of time and/or proportion of income to Christian religious activities as their counterparts in Africa do.

Conversion to Christianity entailed adopting a Western name. This was part of the baptism process. The christening of Africans with Western names was not exclusively an objective of Christianity. It was equally an objective of all European or Western institutions in their dealings with people of African origin. Here, I am reminded of a scene in the highly acclaimed television miniseries on slavery based on Alex Haley's classic, *Roots: The Saga of an American Family* (Haley, 1976). In the scene, a newly-enslaved African who prefers to go by his African name, Kunta Kinte, was forced to assume the Western name, Toby by his slave master. When Kinte refused to accept the Western name the slave master ordered one of the older slaves (the headman) on the plantation to flog him and to continue doing so until Kunta Kinte accepted the name. When this did not work, the slave master ordered the amputation of one of Kinte's feet. The heinous act succeeded in having Kinte to accept the name Toby by which he was thenceforth known.

This episode illustrates the extent to which Western agents were prepared to go to impose Western names on Africans. The question is, why was this a necessary part of the Western colonial and imperial project? To respond to this question with exactitude, it is necessary to first understand the aims of this project. Prominent amongst the project's aims was the supplanting of all African cultures and traditions with Western varieties. There is no question that names constitute an important part of people's culture and tradition. Therefore, replacing African names with European varieties was part of a broader scheme designed to destroy African culture and tradition and simultaneously affirm Europe's or the Western world's perceived cultural superiority over the rest of the world. Apart from serving as a tool of brainwashing, the imposition of Western names destroyed an important aspect of African tradition. Many indigenous African groups were accustomed to androgynous or gender-neutral names prior to the advent of European colonialism and Christianity. For instance, the Banyangi's of Manyu in Cameroon and the Ijagham of Cameroon and Southeastern Nigeria assign female and male children the same names. Thus, names such as Agbor, Arrah, Ayuk, Bessem, Bessong, Egbe, Manyor, Orock, and Tabot, to name just a few, are given to boys and girls alike. Amongst the Baganda of Uganda, children immediately preceding twins are re-assigned a gender-neutral name, Kigongo. Also, in the same culture, the child born immediately after twins is assigned the androgynous name, Kizza. With Christianity and Westernization, these names are often relegated to a tertiary position, as the children become popularly known by their Western names.

My use of the term Western as opposed to Christian names here is deliberate. There is nothing more 'Christian' about the name Stanley, which Christian missionaries were more likely to accept, in comparison to an African name such as Ambe, which they were wont to reject, for the purpose of christening. To be sure, none of the two names appears anywhere in the Bible. Thus, the imposition of Euro-centric names on Africans was part of an effort to 'westernize' as opposed to 'christianize' African people. The imposition of Western names on Africans caused immeasurable, although not readily obvious, damage to African culture and tradition. To appreciate the magnitude of this damage, it is important to first appreciate the fact that one's name constitutes an identity and a window on her culture and self. For the Africans, particularly because of the care with which they name their children, names not only link the bearers to their ancestors but also comprised part of their spirituality.

Another aspect of efforts to 'modernize' and convert Africans to Christianity that has far-reaching implications for African culture and tradition has to do with the languages that were used in institutions of formal education. Here, it is important to note that the record on the use of indigenous languages was exaggerated and distorted. In this regard, some colonial authorities employed the language of favoured or elite indigenous regional groups as the medium of communication during the very early phase of elementary education. For instance, the Germans employed Ewe, the language of the dominant indigenous group in Togo. In a similar vain, the Germans employed Douala, in Cameroon. When the British took over Southern Cameroons, they employed as the vehicle of instruction for the lower grades of elementary school, Douala in the coastal area and Bali (Mungaka) in the grassfield region. These tokenist efforts notwithstanding, it bears noting that colonial authorities strived to produce Africans who, as Aprah (2001) observed, were culturally removed from their indigenous cultures. Kwesi Aprah (2001: 8) noted that,

> The first and most important vehicle for removal and alienation of the educated African from his or her original cultural moorings was the use of the colonial language, English, French or Portuguese.

Colonial and Christian religious authorities not only derided African languages as worthless, but also forbade their use in official colonial business circles. In schools, the use of indigenous languages was deemed a punishable offence. Thus, schoolchildren were constantly flogged for speaking their native languages. 'The acquisition of knowledge was therefore right from the start linked to the use of the colonial languages, . . .' (Aprah, 2001: 8). Members of the indigenous population who could express themselves in these languages occupied a higher point on the social ladder than their counterparts who could not.

To appreciate the extent to which colonial language policies contributed to eroding indigenous African culture, it is necessary to acknowledge the fact that language constitutes more than a medium of communication. A language – any given language – is only one element in a bundle of cultural artifacts. European languages were (are) no different. Readers, novels and other books that were designed to teach colonial languages contained a heavy dose of European history, culture, values, norms and mores, and were essentially subtle tools that were used to supplant African traditional and cultural practices and beliefs with European varieties.

The introduction of Christianity in Africa can also be incriminated as a source of cleavages and hostilities amongst previously unified groups. Relationships between Catholics and Protestants, particularly in Europe, have historically been rancorous. The current situation in Ireland is illustrative. The missionaries representing the various religious institutions were wont to carry their differences to Africa. Accordingly, the peaceful co-existence of Catholic and Protestant missions was a rarity in African villages. If the Catholics were the first to be located in any given village, that village was implicitly out of bounds to Protestants, who must seek alternative location. African Catholic converts were forbidden from intermarrying with their non-Catholic fellow Africans. This effectively inhibited the chances of intermarriages between whole villages. For instance, because the Catholic mission monopolized Njindom Village in Meta, Cameroon, inhabitants of this village could not marry individuals from the other villages in Meta unless these individuals were prepared to convert to Catholicism. It is important to note that Njindom is unique in Meta country where Catholicism is extremely rare. Since the advent of Christianity in Meta, Meta people have increasingly seen themselves less as Meta people and more as either Catholics or Protestants. Consequently, relationships between the residents of Njindom and those of other Meta villages have at best been lukewarm.

Christianity also contributed in no small way to the problem of socio-economic inequalities that is commonplace in contemporary Africa. This problem can be appreciated at two levels, namely the general and the specific. At the general level, Christianity contributed to developing the foundation of socio-economic inequalities between Christians and non-Christians. Institutions of formal education were open exclusively to children of families that had converted to Christianity. Here, it is important to note that Catholic schools were open exclusively to children of Catholic families. Similarly, Protestant schools enrolled only children of Protestant families. Thus, if a village had, say, only a Catholic school, non-Catholic children in that village either had to trek for long distances to other villages to go to school or had to forego school entirely. Christian and colonial government schools were initially not open to girls. Therefore, while boys were given a chance to gain the skills necessary for functioning in modern economic environments, girls were not. For instance, in colonial British Southern Cameroons, there were only three secondary schools, the (Catholic) St. Joseph's College, Sasse, Buea (1938), the Cameroon Protestant College Bali (1948) and the colonial Government Technical College Ombe (1952), and none of them accepted girls. This explains, at least in part, the gender-based socio-economic disparities prevalent in the formal sector of African economies.

Conclusion

The activities of colonial authorities and Christian missionaries were indistinguishable in many domains. Prominent in this regard were the efforts to acculturate Africans. Christian missionaries, like colonial authorities, were bent on supplanting African cultural practices with Euro-centric or Western equivalents. While the reasons for striving to destroy the African's worldview, culture and tradition on the part of the two enterprises, colonialism and Christianity, might have not been the same, they certainly overlapped. For the colonialists, African tradition in many respects, particularly with regards to the ownership and control of factors of production, governance and the administration of public affairs were incompatible with the capitalist manifesto. Christian missionaries considered African belief systems, traditional practices and customs relating to the institution of marriage, the place and role of women in society, production and reproduction, as antithetical to Christian religious doctrine. As noted in this chapter, claims regarding the incompatibility of certain aspects of African culture with Christian doctrine were, in some cases, exaggerated or manufactured. Take the case of polygamy, especially polygyny, for instance. The fact that the Bible does not condemn this practice renders the missionaries' claim indefensible. Efforts on the part of Christian missionaries to exterminate this practice can be best interpreted as part of the broader agenda to Europeanize or westernize Africa by replacing all vestiges of African tradition and culture with Euro-centric substitutes.

Rather early in the colonial era, European colonial authorities and Christian missionaries realized that the African, more than any other human race, defines herself by her culture. In other words, an African sees herself first and foremost, as a Mandingo, a Zulu, a Meta, a Bamileke, a Maasai, or member of any other specific African group. The African thus, becomes less functional and less effective the further removed from her roots she is. With this knowledge, the European was certain that to destroy the African's culture was to render the African powerless. Rendering Africans powerless was necessary to facilitate the success of the colonial project and the infamous slave trade before it.

As I pointed out earlier, the word religion is a derivative of a Latin word meaning ritual. From all indications, there has never been a shortage of rituals in traditional Africa. Therefore, any attempt to associate religion exclusively with Western development is misleading at best. Christianity is simply one of many religions in the world today. Paradoxically, Christianity, especially the Western brand, as opposed to the Orthodox variant, is declining in the West but spreading rapidly throughout Africa. Africans are on the forefront of efforts to spread Western values, culture and customs, including the Christian doctrine in Africa. This is happening at a time when Africa is most in need of strategies capable of propelling socio-economic development on the continent. Yet, to be untrue to themselves as Africans seeking to westernize the continent are, is a prescription for failure. If Japan and other Asian countries have made gigantic strides on the development curve, it is because they have remained true not only to themselves, but also to their culture and tradition. They have ferreted their culture for attributes contained therein that are capable of making a positive contribution to their contemporary development efforts. These countries have not imitated the West simply for the sake of doing so. Rather, they have picked and chosen only the attributes of Western values that can complement their own indigenous values.

Africans will do well to emulate the examples of Asian countries, which have made tremendous socio-economic progress without sacrificing their belief systems, culture and tradition. The blind and uncritical imitation of the West is costing Africans tremendously in socio-economic terms. Consider for instance the fact that Christian holidays are celebrated for longer periods in Africa than anywhere else in the world. To compound this, Africans must contribute large portions of their meager incomes on Church-related activities, some of which I mentioned above. One cannot but imagine the number of socio-economic development projects that Africans must forego in order to meet their obligations with Western institutions. Another aspect of African culture that has suffered significantly as a consequence of westernization efforts in Africa relates to the social construction of gender, which manifests itself in many ways, particularly the assignment of gender-specific names to children and the act of viewing the male gender as the pre-ordained superior to the female gender.

This view has been propagated for thousands of years, especially by the Bible, which holds that God created Eve from one of Adam's right rib bone. The implications of this doctrine have been far-reaching in Africa. As I show in subsequent chapters, the disempowerment of women in Africa occurred largely following the European conquest and the concomitant introduction of Euro-centric values and especially those regarding gender relations. Although most parts of Africa prior to the European conquest were patrilineal, they were not necessarily patriarchal.

Chapter 4

The Traditional African Family

Introduction

Eminent Africanist, Ali Mazrui is accurate when he asserts that the African family remains an enigma to many a non-African. It has demonstrated remarkable resilience by surviving centuries of ferocious assault. The African family continues to maintain its unique structure and identity despite several centuries of brutal assault from exogenous forces. The fact that the African family has maintained its traditional identity and raison d'être has surprised even the most stoic Western observer or change agent. However, despite constituting the focus of attention especially during the colonial era, the African family remains one of the least understood and often the most misrepresented of all indigenous African institutions. The family is a highly valued social unit, which plays an extremely important role in African society. African tradition requires people to have a high sense of obligation to their kin in particular and their community at large. The importance of family in traditional Africa, where there are no formal social welfare institutions, derives from the fact that it provides a platform on which members offer and receive assistance, encouragement and advice (Obijiofor, n.d.). The extended African family thus, amongst other things, plays the role that social welfare institutions play in the West.

This chapter seeks to shed some light on this very important but least understood institution. Of particular interest are questions relating to the 'how's, what's and why's' of certain attributes of the traditional African family that have become the subject of rancorous debate in political, social and economic circles. The following aspects of African culture and tradition that are of relevance to the discourse on the family in African society will be explored.

- Kinship and the extended family phenomenon;
- Marriage and especially, polygamous marriages,
- Extended periods of lactation;
- The inheritance clause; and
- Rights of passage.

Additionally, the chapter identifies and discusses some of the major strategies that were employed by colonial and Christian authorities to eradicate or alter the traditional African family.

Kinship and the extended family phenomenon

Kinship has to do with relationships based on blood or marital ties. In other words, people may be regarded as kin if they are related by blood – that is, a consanguine relationship – or through marriage. Both relationships are considered very important in the traditional African context. In Africa, as opposed to most parts of the world, whole families and not individuals, marry. In other words, marriage is viewed as an alliance between two kin groups as opposed to a contract between two individuals, as it is typically the case in Western societies. The alliance often begins with an arranged matrimonial relationship between a young man and a young woman in the two families.

The bridegroom's family is always required to meet the cost of the bridewealth. Until the advent of colonialism and the concomitant monetarization of the African economy, bridewealth usually assumed the form of cattle and other non-monetary items. In this latter regard, the bridegroom was typically required to perform an assortment of chores, including but limited to clearing farmland and fetching fuel wood for the future mother-in-law. The bridewealth serve as a token to demonstrate that the groom and his family are interested in having the bride as a member of their family. Furthermore, it serves as a testament to the preparedness on the part of the groom's family to take utmost care of the bride throughout her entire life. The need for this commitment is magnified several folds once we take into account the fact that divorce is seriously frowned at and in fact, treated disdainfully in traditional Africa.

Khapoya (1998) identifies four additional reasons for the bridewealth. First, it is meant as an expression of appreciation by the groom's, to the bride's, family for a praiseworthy job in raising a marriageable young woman. Second, bridewealth serves as a modest compensation to the bride's family for foregoing her real and potential productivity. This is because upon marriage, the productive potential or labour power of the bride is automatically transferred to the groom's kin. Third, the bridewealth, guarantees the rights of the groom and his family over all children brought forth by the bride during the course of the marriage. Thus, the reproductive potential of the bride effectively falls under the control of the groom and his kin upon marriage. Finally, bridewealth is designed to cement the bond between the bride's and groom's families or kin.

The decision regarding future marital partners does not reside exclusively in the hands of the bride and groom. Rather, this decision usually falls in large part in the hands of their families. This is a testament to my earlier assertion that marriage in traditional Africa is an alliance between two families as opposed to a contract between the bride and the groom. Thus, in traditional Africa, even in matters such as marriage that Westerners typically view as personal, the interest of the individual is ultimately subordinated to that of the collective. This custom was not only interrupted, but it was also undermined during the colonial era when European colonial authorities and Western change agents strived to assimilate and acculturate Africans by all means necessary.

In Vincent Khapoya's (1998) praiseworthy work on 'the African Experience,' he sheds light on the notion of kinship in African society. As Khapoya argues, and I concur, marriage acquires a significantly greater importance in traditional African society. There are several reasons for this. Prominent amongst these reasons is the fact traditional Africa is by definition agrarian.

This concept of family differs sharply from what obtains in Western societies. For one thing, traditional Africans practice polygamy. I return to the issue of polygamy when I discuss the different forms of marriages practiced in traditional African society later in this chapter. For now, suffice to state that polygamy may assume one of two forms, polygyny – one man marrying more than one wife – and polyandry – one woman marrying more than one husband. The former is the more commonplace. For another thing, the concept of family in traditional African society extends beyond the nuclear family – mother, father and children – which is more popular in Western societies.

The traditional African family typically includes all individuals belonging to a common lineage. Most African languages do not contain equivalents for words such as half brother, half sister, niece, nephew and cousin that are commonly found in Western languages. Rather, the word brother or sister is used in almost all African languages to refer to one's half brothers, half sisters, and nephews, nieces and cousins. One's sisters' and brothers' children are referred to as one's children. The word uncle refers only to one's maternal uncles. However, one's paternal uncles are one's 'fathers.'

Westerners are wont to construe this apparent lack of specificity in reference to consanguine relatives as testament to what they believe to be a lack of versatility on the part of African languages. Yet, nothing could be further from the truth. The deliberate use of less precise terms in reference to one's kin is meant to accentuate the importance of consanguine relationships in traditional African society. Africans have always been aware of the need for family members to live in unity. In his discussion of the traditional African family, Tembo (On-Line) suggested, that Europeans were particularly baffled by this aspect of African tradition – that which had to do with distinguishing amongst family members. To demonstrate the fact that Africans were never interested in compartmentalizing or identifying family members with the degree of specificity common amongst Westerners, he cites the case of the Baganda of Uganda. Here, children refer to their father's wives as their mothers and their paternal uncles, as their fathers. These appellations, as Tembo hastens to note, and I concur based on my experience with the Meta and other cultures of Cameroon,

> carried with them all the heavy social obligations demanded of a mother or father, daughter or son. There was never a distinction between the biological and non-biological kin as far as primary parental obligations were concerned.

This is a function of the fact that Africans have always had to wrestle with adversity throughout history. Prominent amongst these adversities is the relatively short life expectancy of Africans. Thus, it is common for children to be without both biological parents and even more common to be without one of them.

African tradition requires that upon the death of a man without adult male children, one of his brothers must immediately assume full charge over his widow or widows and their children. Thus, the surviving brother is required to become the 'husband' to the widow(s) and 'father' to the 'semi-orphans.' There are several similarities to this situation in the Bible. For instance, according to Deuteronomy, 25: 5–10, a woman must not remarry outside the family. In the event that the man is survived by adult male children, one of the males (usually, but not always, the eldest male offspring), designated in the man's will as his heir, is required to take charge of the whole family. In this case, the heir must take over his late father's estate and assume the role of husband to his late father's widows (excluding, of course, his own mother), and must function as a 'father' to his siblings, including his half-brothers and half-sisters.

Upon the death of a married woman, her co-spouses assume the responsibility of raising her children. In most African cultures, children born to unmarried women are considered members of her paternal family. Thus, the illegitimate child phenomenon is non-existent in traditional African society. Here, the grand parents typically raise the children of their young unmarried daughters. Under these circumstances, the children are socialized to treat their grand parents as if they were their biological parents. In such cases, the child is considered part of the mother's paternal lineage. The dominance of one's paternal side of the family as suggested here is a function of the fact that most traditional African societies have patrilineal descent systems. The relevant literature suggests that descent systems can be meaningfully discussed under three major categories (Khapoya, 1998). The first and most common is the patrilineal system, wherein one's descent is traced through one's father and his male ancestors. The second is known as matrilineal descent in which ancestry is traced through one's mother's forebears. The third is called duolineal or bilateral descent in which one's ancestry is traced through both parents. The converse of this, unilineal descent, is the system in which descent is traced through one side – the mother's or father's side – of the family. In patrilineal systems, which are the most dominant in Africa, a young woman automatically becomes a member of the husband's family in particular, and community as a whole, upon marriage.

Occasions such as marriage, birth and death are given a lot more attention in traditional Africa than in the Western world. In a majority of African cultures, marriage is not permitted between people who share a common ancestry. Furthermore, blood relatives can neither be married nor be romantically involved, although the precise degree of consanguity does however differ from one group to the next. It is believed that breaches of these customs ultimately result in misfortune. This misfortune can only be prevented or reversed through certain rituals. Such breaches are considered more serious when a romantic relationship between members of the same lineage results in pregnancy and subsequently childbirth. The misfortune in this case is believed to be more severe and may sometimes result in the death of the resultant child.

Types of marriages

The various forms of marriage found in Africa can be discussed under five broad categories, namely polygamous, sororate, levirate, surrogate and ghost marriages.

Polygamy

In the mind of most persons, the word polygamy often conjures the notion of one man marrying two or more wives. However, that is only one form of polygamy known as polygyny. Polygyny is commonplace throughout Africa. As Tembo (On-Line) notes 'the polygynous joint family, consisting of a man, his wives, and their children, is the ideal for most Africans.' In fact, most studies 'conducted from the 1930s to 1950s indicate that polygyny was common virtually in all regions of Africa' (Tembo, On-Line).

The other variant of polygamy is known as polyandry. Polyandry entails one woman wedded to two or more men. This form of marriage is uncommon but not, as some (e.g., Khapoya, 1998) have suggested, non-existent in Africa. Polyandry, which may be fraternal – that is, when two or more brothers are married to one woman – or non-fraternal, can be found among some cattle-rearing groups in East and southern Africa, among the Banyabkole and kindred tribes of northeastern Africa, northern Nigeria and northern Cameroon (see Levine and Sagree, 1980).

Four other, albeit less common forms of marriage can be found on the continent. The first of these is known as sororate marriage and entails a man marrying his wife's sister. The second is known as levirate, which characterizes a situation in which a man inherits his dead brother's wife. The third is referred to as surrogate marriage, which involves a woman 'marrying' another woman. Here, I hasten to note that this bears no resemblance to the genre of homosexual liaisons that have constituted the subject of socio-political debates in the U.S. and other Western countries in the recent past. Rather, a levirate marriage usually occurs when a man dies without any male offspring and one of his daughters inherits his estate. In such a case, it is customary for the female heir not to move out of the family compound. Rather, she must seek to have children that will keep the family name alive. Doing so often entails 'marrying' one or more young women with the goal of bearing more children in the deceased man's household. In this case, the young woman regards the (female) heir to the deceased man's estate as her 'husband.' The heir in turn is required to conduct all the necessary formalities, including the payment of the bride price as well as providing for the young woman just as a husband would.

Finally, there is what anthropologists have characterized as 'ghost marriage.' Ghost marriages typically occur between a woman and a dead childless man. The intention in such cases is usually to ensure that the dead person has a 'heir' to sustain the family name.

Wife-swapping

Before leaving the issue of marriage, it is necessary to draw attention to the issue of wife-swapping. This practice is extremely rare and may be non-existent in contemporary Africa. However, it was not that uncommon in pre-colonial Africa, as Macdonald (1890) observed among some South African natives. Here, men occasionally exchanged their wives. However, this arrangement was usually of a temporary nature, as each woman involved remained the wife of the original husband. Also, and most importantly, any children born during the exchange belonged to the woman's original husband.

The occasion of childbirth

As stated earlier, the arrival of a newborn is considered a very important occasion marked by a series of elaborate rituals in traditional African society. These rituals differ from culture to culture. Macdonald (1890) describes this occasion, as celebrated amongst natives of the southern African region. In this region, a woman who has just given birth is secluded for one month (in lunar terms). This period of seclusion is known as incubation and is deemed necessary to maintain the nursing mother's fertility. It is believed that failure to observe this period would result in infertility.

Once a child is born, and the wise woman reports favourably, the child's father is required to slaughter an animal, usually a sheep, a goat, or an ox. This is meant as a token of appreciation on the part of the newborn's father to the spirits of his ancestors and household gods. It is believed that failure to make this sacrifice would result in the risk of death to the child during the first few years of life.

> Neither father nor mother can transfer children to another clan, and persons residing among other tribes than their own still regard themselves as members of that to which forefathers belonged' (Macdonald, 1890: 267).

The rearing of a child is the responsibility of all members of that child's extended family in particular and the village to which she belongs as a whole. Thus, the all-too-familiar African adage that, 'it takes a village to raise a child.' In practice, this means that doing everything to ensure that the child develops into a productive member of society is the responsibility of the entire village. Thus, every member of the child's village has the right to discipline the child should she step out of line. Children, particularly infants, are considered especially vulnerable. It is believed that evil spirits can seize infants. Thus, throughout traditional African societies, extraordinary care is taken to protect children. For instance, to divert the attention of evil spirits from the child, and as a means of protection and beautification, children in Himba, Namibia, are adorned with jewelry and smeared with a mixture of red ocher from the earth and animal fat at the tender age of only a few days.

Breastfeeding lasted at least two years in pre-colonial Africa. Children were at least three years apart. It was considered a taboo for a nursing mother to indulge in sexual activities. Thus women observed a strict code of abstinence from all sexual activities during lactation. Traditional Africans believed that sexual intercourse during lactation could lead to the contamination of breast milk, which would invariably cause the baby to become ill and possibly die. It was equally a taboo for a woman to indulge in such activities during menstruation and pregnancy.

Anthropologists working in colonial Africa were puzzled by the fact that the spacing of children was done with exactitude in pre-colonial Africa, which had no 'scientific' or so-called 'modern' concept of time. Long ago, Africans had devised alternative means of knowing exactly when to wean an infant off breast milk. One example is illustrative. The Metas of North West Province, Cameroon, based their judgment about when to wean a baby from breast milk on the baby's physical abilities. In this regard, a baby was considered ready for weaning once she was able to cross the doorsill or doorstep. This threshold is usually about 45 cm high. As I noted elsewhere (Njoh, 1999), when an infant is old enough to scale the 45 cm wall in and out of the building, it is interpreted as a sign that she can be weaned from breast milk.

Child adoption is commonplace in traditional African society, but not in the same sense as it occurs in the West. A family with means may raise the children of poor homes. In Africa, the rich have an obligation to help the poor. It may be here noted that this practice bears a striking resemblance to what obtains under the 'poor law' in some Western countries such as Great Britain. Amongst the natives of the predominantly cattle-rearing southern African region, a wealthy person is expected to lend his milking cow to a poor neighbour with young children. Failure to do so would result in disgrace to his name and dishonour to the chief of his clan and ultimately a severe punishment by fine or confiscation of goods by the tribal council.

Transition into womanhood

Quite arguably the most important stage in the life of a child in traditional Africa is puberty. It is therefore hardly any wonder that this stage is marked by a series of elaborate ritualistic ceremonies in most parts of the continent. These ceremonies are connected with initiation and are designed to remind the initiates that they have attained an important milestone in their lives. The girls are reminded that they are, from that point onward, women and are therefore expected to do 'womanly chores' and in general, conduct themselves as 'women' and not 'girls.' The actual transition into adulthood for girls is usually preceded by a period of serious training designed to prepare them for marriage. In some traditional African cultures, girls who have attained puberty but are unmarried are required to adorn themselves in a distinct manner that makes their status clear to potential suitors. For instance, in traditional Himba society in Namibia, unmarried girls were required to wear a rosette of goatskin in their hair and a beaded coil necklace to signify their marital status as 'single' and 'available.'

The initiation period for any given cohort of girls often begins once the girls attain puberty and culminates in a public ceremony designed to 'release' them as women into the world. This process differs slightly from group to group. The following specific cases culled from a number of major regions across the continent are illustrative and serve to shed more light on this elaborate process (African ceremonies, On-Line).

Shai initiation, Ghana

As part of Shai 'outdooring' or initiation to womanhood, girls are required to display their beauty, elegance and grace by putting on *Dipo-pe* straw hats, and performing the *klama* dance. The dance emphasizes the graceful movement of hands and feet. The hats accentuate quiet elegance when the girls turn their heads demurely downward. The occasion provides suitors (often constituting part of the watching crowd) an opportunity to select future brides.

Krobo initiation

The Krobo, also a Ghanaian subgroup, require young women to undergo a three-week intensive initiation ritual to mark the young women's maturity as young adults. The ritual, which is particularly designed to sever all ties the young girls have with childhood, involves instructions on what it means to be a woman, how a woman needs to conduct herself in the public and domestic spheres and so on. The initiation period usually culminates in a series of outdoor activities prominent amongst which is an occasion wherein the girls display their feminine grace, beauty and dancing skills to the chief, relatives, and most importantly, potential suitors.

Turkana, Kenya

In traditional Turkana society (Northwestern Kenya) unmarried girls were required to wear a multilayered beaded necklace with three back pendants to indicate their marital status as 'unmarried' and 'available.'

Maasai, Kenya

A long-standing Maasai ritual forbids brides making the first official and usually long journey to their marital home to look behind lest they turn to a stone. Maasai Warriors, known for their bravery, are highly revered in their local communities and feared by their foes, have almost unrestricted access to young local girls. In traditional Maasai society, the young girls who accompanied (or accompany) Maasai Warriors take advantage of their pre-marital status to become intimately involved with multiple (up to three) partners. Each female companion of the Maasai Warriors may have as many as three of the Warriors as lovers. The lovers in any of the young women's lives are hierarchically ordered from the most important (the sweetheart) to the least important. Each young woman makes the hierarchical order abundantly clear to the men who must respect each other and refrain from any behaviour that may be indicative of jealousy.

Meta, Cameroon

The Maasai culture that appears to condone pre-marital sexual relations as suggested above, contrasts sharply with the culture of pre-colonial Meta society in Cameroon. In pre-colonial Meta, a gift basket containing several valuable items, principal among which is a large cylindrical block of dried salt in leaf wrapping was customarily sent to the young bride's mother on the day following the bride and groom's first night together. In the event that bride was found to be a virgin, the block of salt was sent in its entirety. However, if the young bride had lost her virginity previously, a large hole was drilled through the block of dried salt prior to sending it to her mother. A woman who received a post-wedding night gift basket containing a complete block of dried salt was extolled and respected in the community for a wonderful job raising a disciplined bride. On the contrary, a woman who received a hollow salt was shamed for having done a poor job in this regard.

Transition into manhood

The transition from boyhood into manhood involves several steps. Prominent in this regard in most African customs is the circumcision of the young initiate. In pre-colonial Africa, his typically occurred between the age of sixteen and eighteen. Circumcision was, and in some areas, arguably the most important of all the rites marking a boy's transition into manhood. It is believed that manhood is attained through the shedding of blood during circumcision.

In some cultures, this process usually involves undisguised and occasionally, thinly veiled sexist overtones. For instance, during the actual removal of the foreskin on the penis, the young initiates are told that 'the spear has been sharpened' and that it is now ready for use in striking the 'animal.' Here, the 'spear' alludes to the circumcised penis while the 'animal' is in reference to women. The obvious vulgarity here is unique and quite unusual as sexual organs are seldom mentioned in African languages.

The male initiation process amongst the natives of Southern Africa particularly fascinated the Reverend James Macdonald. He described this elaborate process in his 1890 piece, 'Manners, Customs, Superstitions, and Religions of South African Tribes.' Circumcision in this region is conducted during a specific season of the year, when crops are just beginning to show signs of ripening. The process usually begins with the village doctor or medicine man circumcising all the young men of the village who have just turned sixteen or who turned sixteen subsequent to the last circumcision event. After circumcision, the initiates are isolated in previously prepared residential units. These units are usually located away from the rest of the population. During the period of seclusion, the initiates are closely monitored and supervised by a team of appointed men. The neophytes are prevented from indulging in potentially distracting activities, particularly sexual intercourse. They are daubed all over with pure white clay. This white clay constitutes the badge of identification for the young men during their probation. During this period, the 'white boys,' so called because of the colour of the clay used in daubing their bodies, are expected to toil for their own food. Success in this regard is often extolled while failure is often met with retribution in the form of severe corporal punishment. The neophytes are also required, as part of this initiation-related probation to perform serious physical exercises, including, but not limited to, dancing and running. Additionally, the initiates 'are often kept awake for several consecutive nights; beaten over the arms and thighs; and deprived of food for varying periods' (Macdonald, 1890: 268). The purpose of these formalities is to harden the initiates and prepare them for the difficulties they are likely to encounter in life. One testament to the probation's rigour is the fact that some of the initiates die during the process. At the close of the probation, the initiates wash off the white clay from their bodies and are presented with a blanket each. Also, they burn all the tools and utensils that were used during the probation. Then, they proceed to the residence of the head of the clan. It is at the village chief's or clan leader's compound that a special group of local elders anoint the young initiates with oil and smear them with red clay (ochre).

At this point in the initiation process, the neophytes listen to speeches from the elders of the clan. A common thread that usually runs through all the speeches is the message that they (the initiates) are no longer boys and that they will never be boys again. The burning of the utensils and tools they had used during probation, the initiates are told, signifies that the tools, utensils and other things in their lives up to that point, belonged to their boyhood and must never re-appear at any point in the rest of their lives. As the men that they have become, they were now expected to do 'manly things,' such as bearing arms, going to war to defend their chief, avenge his wrongs or fight to protect their communal land and other property. The menial chores such as herding cattle, hoeing and similar tasks must from then onward, be executed by their younger siblings.

Age-grades and socialization

In traditional Africa, individuals who undergo the initiation process at the same time usually belong to a common age-grade. An age-grade is a system of social organization whose membership is based on age cohorts. Age-grades were common in pre-colonial Africa and continue to exist in one form or another throughout the continent today. The roots of this system go deep into African history. It constituted the foundation of the ancient African educational system, which began in Nubia-Kush, Khemet (Egypt) and was later adopted in Abyssinia from where it spread to other parts of Africa several centuries B.C. This system is believed to have influenced and inspired the educational structure of the Greeks, Mesopotamians, Hebrews, East

Indians and the Chou Dynasty in China, which took over from the Black Shang Dynasty about 1100 B.C. (Black College/University Student, On-Line).

The age-grade system takes advantage of a universal fact, namely the ageing process. In traditional Africa, age is accorded a priority place in society. The older one gets, the more respect one commands. An age-grade thus denotes one position on what is generally a rigid social ladder based on age. In ancient Africa, people were divided into groups according to their ages. Each age group was then taught skills deemed apropos and necessary for survival and socio-economic functioning at that age group. The time span between age-grades is not clear. Among the Sotho and Tswana of southern Africa, for instance, young boys and girls are recruited into age sets or age grades every four to seven years.

Three age grades, child, junior and elder are typical of most groups in Africa although it must be noted that these divisions often tend to vary and nuance from one region to another. For instance, the Karimojong of north-eastern Uganda emphasized only a single division between seniors and juniors. The Nuer of southern Sudan, also classify males into two main groups, senior and junior, based on their initiation status. Among the Arusha of northern Tanzania, the two major grades, senior and junior are further divided into older and younger subdivisions. The Nupe's of Nigeria recognize three age grades.

As each group in traditional Africa advances in age, its members are taught new and progressively advanced and more sophisticated skills. Furthermore, societal roles are assigned to different age groups based on their abilities and capabilities. Among the Maasai and other groups known for their warring proclivities, young men who have just been recently initiated into manhood are assigned the role of warrior. Men who have just attained manhood are particularly believed to be physically strong and aggressive hence, considered most suitable for this role. Members of the older cohort or age groups are considered to be seasoned, endowed with the genre of wisdom that comes with age, and therefore assigned the role of social, political and moral leaders. In pre-colonial West Africa, members of younger age groups (12 to 30 years of age) were assigned tasks such as constructing and maintaining public infrastructure. In pre-colonial Benin, they were assigned the additional task of repairing shrines or council houses, as well as carrying the tribute of yams and palm oil to the King (Oba).

As Bennett (1998) notes, each status is important in its own unique way as it entails a different set of rights, responsibilities, duties and privileges in relation to other statuses in the system. The differences are typically reinforced by a set of rules that specify the behaviour commensurate with each status. Age-grade systems facilitated political control in pre-colonial Africa, particularly because they transcend family, kinship or other ties. This made it possible for senior rulers to effectuate command and control over otherwise recalcitrant members of the junior age grades. Amongs the Nandi of western Kenya the recruitment of members into age grades occurs every 15 years.

African concept of family, development implications

Colonial authorities and Christian missionaries viewed African belief systems, culture and traditional practices as antithetical to their development efforts. To them, African customs were unquestionably incompatible with efforts designed to realize colonial development projects. This was particularly true with respect to African customs having to do with the related institutions and/or activities of family, kinship, human development and socialization.

African culture and traditional practices are not antithetical to development efforts. Rather, the changes introduced by European colonial authorities, missionaries and other Western change agents sowed the seed for the nagging development problems characteristic of the continent. In the remainder of this chapter, I marshal evidence to demonstrate that African tradition and cultural practices would have proved more beneficial to contemporary development efforts in Africa were it not for the so-called modernization efforts that were initiated during the colonial era. I focus on the following specific domains.

- The family as a unit of economic production;
- Polygamy and wife inheritance;
- bride price;
- Inter-birth intervals; and
- Initiation and age-grading.

Family as unit of economic production

The traditional African family has always constituted the basic unit of economic production. For instance, a typical traditional African family, comprising a man, his wife or wives and children, jointly worked on the family farm and lived in the family compound. However, with the advent of colonialism and the concomitant changes, these traditional structures were effectively destabilized. Two specific developments in this regard are particularly noteworthy. The first has to do with the introduction of cash crop agriculture. Colonial authorities introduced cash crops and plantation agriculture in sub-Saharan Africa. Only men were assigned the meaningful roles in this area. Accordingly, women were left to single-handedly take charge of food crop cultivation. The formal economic sector also favoured men, as they were the only ones who were employed in the colonial government service, while women were relegated to the role of preparing and selling food to colonial government and other formal sector workers. This sex-typed division of labour laid the foundation for the gender-based socio-economic disparities prevalent in contemporary sub-Saharan African economies. The second economic development with serious negative consequences for African families was the introduction of large-scale commercial mining. This activity, like other colonial economic development projects, had a propensity for hiring exclusively men. Thus, men were often compelled to be away from their families and in remote mining locales for extended periods. Residential facilities in mining locales made no provision for families. Rather, these facilities were typically of the single-room dormitory genre, equipped with communal toilets and kitchens. One consequence of this protracted confinement of men was the growth and proliferation of prostitution in the vicinity of mines and plantations.

The growth of prostitution has also been linked to the colonial economic development policies that blatantly discriminated against women and ignored the importance of traditional African families as well as other customary institutions. For instance, in colonial Kenya, colonial economic and social development policies contributed significantly to a rise in the rate of prostitution (White, 1990). According to Louise White (1990), the brand of prostitution that evolved in colonial Kenya, especially in the colonial government capital of Nairobi is worthy of note as it posed a real threat to the institution of marriage throughout the colony.

Polygamy and wife inheritance

Missionaries persuaded Africans to convert to Christianity and abandon their indigenous cultures, traditions, customs and beliefs. Colonial authorities aided the process by setting up systems and creating laws that were designed to promote Christian/Western ideals. To accomplish this objective, it was necessary to supplant several traditional African values. European colonial authorities and Christian missionaries found the African tradition of polygamy especially contemptuous and were therefore bent on replacing it with monogamy. Missionaries preached that polygamy was against Biblical teachings. This was despite the fact that several prominent figures in the Bible, including but not limited to Isaac, Jacob, and Abraham practiced polygyny. In fact, Christianity never prohibited polygamy until it was Europeanized a few centuries ago. Thus, apart from prominent Biblical figures such as prophets, Christians practiced polygamy for a long time during their history.

When the attention of missionaries was drawn to this fact, which unequivocally contradicted their preaching, they countered by arguing that polygamy was an anachronistic practice that occurred exclusively in the Old Testament. The New Testament, the missionaries contended, condemned polygamy. However, it is worth noting that nowhere in the New Testament is polygamy or any of its variants condemned. To be sure, 1 Timothy 3: 12, quite possibly the only word on how many wives a Christian may marry, states thus, 'Let deacons be the husbands of one wife . . . .' Here, it is clear that not everyone was being instructed to have only one spouse. Rather, the message was specifically directed to deacons. One does not have to peruse the Bible to realize that the instructions of the missionaries were out of touch with reality. Note that the missionaries originated from the Western world, where remarriage was (is) permitted subsequent to a divorce or death of a spouse. In a way, this can be seen as serial polygamy – in other words, an individual marrying multiple spouses, one at a time during the course of his/her life. To put this in perspective, it is important to note that in Ancient Greece, a woman was forbidden from remarrying subsequent to the death of her husband.

Efforts to make monogamy the norm in Africa had nothing to do with true Christian teachings. Rather, these efforts were part of a larger project to impose European cultural ideals on Africans. As in other domains, the missionaries' strictures against polygyny were the product of a Western mindset rather than an adherence to Biblical teachings. Missionaries, colonial authorities and other Western change agents attempted to conceal their real intentions by arguing that polygyny constituted a means for the men to realize their sexual fantasies, exploit and subjugate women.

However, because there was also the possibility of one woman marring multiple men in a few African cultures, this argument is indefensible. Polyandry, the practice of a woman marrying more than one man is of two varieties, namely fraternal and non-fraternal. Fraternal polyandry connotes the practice of two or more brothers marrying one wife. Non-fraternal polyandry has to do with a woman marrying two or more non-consanguinely related men.

There are a number of distinct advantages associated with fraternal polyandry, especially for agricultural societies such as those found in Africa. First, by making it possible for two or more brothers to marry one woman, polyandry becomes a means for maximizing the utility of scarce resources. Some of these resources may be indivisible or are likely to be less productive when divided up. In the case of family farms, polyandry ensures that they are kept in one block, thereby maximizing their productivity. Second, African tradition in most parts of the continent dictates that when a man dies, one of his brothers must marry the man's widow and raise seed in his name. In this case, polyandry makes it possible for the widow and her children to be cared for. Thus, closely related to this is a third advantage of polyandry, namely the fact that it provides social and economic security for widows.

Furthermore, it can be argued that polyandry helps to reduce family tensions that may arise over conflicting claims to family-owned property. Perhaps more importantly, polyandry reduces the incidence of widowhood and its concomitant problems. This is because of the unlikelihood that all the brothers jointly married to any given woman will die simultaneously.

As mentioned above, polyandry may also involve husbands exchanging their wives. To be sure, as suggested earlier, the objective in this case has nothing to do with the apparent perverted wife-swapping orgies that go on in some quarters in Western society. Rather, such relationships constituted a strategy designed to deal with health problem – the problem of infertility. Once we understand the importance of children in the lives of Africans in particular and humans in general, it becomes easy to appreciate polyandry as a strategy for combating the problem of infertility. In short, this arrangement made it possible for otherwise childless couples to have children. In contemporary Western societies this need is met by fertility clinics or fulfilled through adoption centres.

The other form of polygamy – in fact, the most commonly discussed one – is polygyny, which entails one man marrying more than one wife. Khapoya (1998) has identified and discussed some of the main advantages of this form of marriage. To fully understand this advantages or justifications for polygyny, one must first appreciate the importance of marriage in traditional Africa. Here, marriage has always been considered as socially desirable. This is not unlike the case in Western societies where there is an overt bias against unmarried persons. For instance, lending institutions charge unmarried persons interest rates that are significantly higher than those charged to married individuals. Similarly, unmarried persons pay more for automobile insurance than their married counterparts. To the extent that marriage is socially desirable in Africa, and there is an undersupply of marriageable men, polygyny then becomes an effective mechanism for guaranteeing marriage to the surplus marriageable women in society. The undersupply of marriageable men or men in general resulted from several factors. The most prominent is the fact that infant mortalities rates were higher among boys than among girls. Thus, polygyny can be seen as a strategy that is socially necessary not only to ensure the continuation of society, but also provide for needs of the many marriageable women who would otherwise not enjoy the status and benefits concomitant with marriage.

Second, polygyny serves as a strategy for economic development in agrarian societies. Such societies usually require a large number of field hands. Thus polygyny affords otherwise small families a means to rapidly augment their numbers. However, it is important to avoid the pitfall of erroneously assuming that women in polygamous relations have a higher rate of fertility than those in monogamous households.

Third, there is the fact, as alluded to above, that male children experience a lower survival rate than their female counterparts. Yet, because most African societies are patrilineal, male children tend to be more highly sought after than females. Boys are usually, although not always, the heirs to their fathers. Thus, multiple wives significantly increases a man's chances of having a male offspring.

Fourth, polygyny ensured that a widow (and her children) could be inherited and cared for by her brother-in-law even if he was already married. Fifth, polygyny constitutes a source of wealth and social prestige in traditional Africa. The case for polygyny in this case draws its inspiration from the fact that polygamous families are (were) almost always wealthier than the monogamous ones. Here, there is a chicken-and-egg conundrum as it is not quite obvious as to which factor precedes – the wealth or polygamy? The conundrum becomes more glaring once we consider the fact that bridewealth usually entails huge expenses thus, making polygyny an expensive undertaking, which only the wealthier men can afford.

Sixth, polygyny guaranteed men an alternative source of sexual gratification during a wife's pregnancy and lactation (usually no less than two years) following the birth of the child. The tradition of most African groups forbids sexual intercourse during pregnancy and during breastfeeding.

Seventh, polygyny serves as an effective strategy to control population explosion. Contrary to popular belief, women in polygynous households experience a level of fertility that is considerably lower than their counterparts in monogamous marital relations. Studies of polygyny among the Negev Bedouins of Southern Israel (Muhsam, 1956), the Temne of Sierra Leone (Dorjan, 1959), and the Kipsigis of Kenya (Mulder, 1989) revealed that polygynously married women had fewer children than their monogamous counterparts. The studies further suggested that women whose husbands have more to three wives experienced levels of fertility that were significantly inferior to those of women whose husbands had one or two wives.

Bride price

It is easy to understand why some Africanists (e.g., Tembo, On-Line) abhor the use of the term bride price in reference to the valuables that are transferred from the groom and/or his parents to the parents of the bride in traditional Africa. While this transfer of valuables constitutes a prerequisite for legitimizing the matrimonial union between the bride and the groom, there is no justification for alluding to the custom as 'purchasing' or 'buying' a wife. Although opportunistic parents, especially fathers, have been inclined to abusing this custom in contemporary Africa, traditional Africans did not receive anything in excess of what tradition dictated as necessary testament to the potential groom's desire and willingness to wed the potential bride.

The bride price has several advantages. Prominent in this regard, and as suggested above, is the fact that it serves to attest to the groom's willingness and preparedness to enter what is by all standards a very serious relationship, namely marriage. Second, it is a means of thanking the parents of the potential bride for raising a marriageable young lady. Third, it is a symbolic attempt to compensate the bride's family for the real and potential economic contribution of the bride that they must lose upon marriage. Most African cultures are patrilineal. Thus, upon marriage, a woman automatically becomes a member of the husband's family. Fourth, bride price serves as an 'insurance policy' for the bride. Bride price in most parts of traditional Africa includes livestock, particularly cattle. In most cases, the bride is not permitted to join the groom until the (female) cattle (e.g., cows, goats, etc.) that comprise part of the bride price is already in the hands of the bride's family. Once there, it is expected that the cattle will be raised to multiply as much as possible. In the unfortunate event of divorce, the bride is sure of returning to her parents' home and relying on the resultant herd of cattle for economic sustenance.

Inter-birth intervals

As stated above, traditional Africans observed longer inter-birth intervals. Infants were breast-fed for an average of three years. The observance of postpartum abstinence meant that during this period the nursing mother had no conjugal relations whatsoever. Traditional Africans believed that sexual intercourse during breastfeeding had the potential of contaminating breast milk, which in turn could threaten the health of the infant. Thus, children were typically born at an interval of at least three years apart. Colonial authorities, Christian missionaries and other Western change agents considered African beliefs and concomitant practices in this, as in other regards, as irrational and absurd. Two leading authorities on female fertility and reproductivity in colonial Africa, Doctors Trolli and van Nitsen (quoted in Hunt, 1988: 410) commenting on traditional African customary birth spacing practices during the heydays of the colonial era stated thus:

the present situation is irrational. Sometimes women breastfeed during three years. In the course of the approximate thirty years during which women are susceptible of becoming mothers, to place periods of three to four years during which they can have only a single child, while nature would certainly permit them to support more frequent pregnancies without harm.

Colonial authorities and other Western change agents were particularly perturbed by the fact that over the course of a typical African woman's life, she brought forth significantly less children than she would have if shorter inter-birth intervals were observed. Their perturbation was borne of their desire to significantly increase the productivity of the colonial economy. Colonial authorities were highly desirous of several energetic and productive hands in the colonial mines and plantations. Traditional African reproductive practices as they were, sharply conflicted with this particular goal of the colonial project. Accordingly, colonial authorities decided in favour of initiating efforts designed to increase the continent's population.

Noteworthy in this regard are the efforts of the Belgian colonial authorities in the Democratic Republic of Congo or what at the time was known as the Belgian Congo (Hunt, 1988). Here, the colonial authorities realized rather early during the colonial era that 'a sufficient population was an absolute necessity for the colony's 'harmonious development'. . . without black labor, our colony would never be able to send to Europe the wealth buried in its soil' (Hunt, 1988: 404–405). Accordingly, the Belgian colonial authorities enlisted the efforts of other Western agents of change, particularly the *Ligue pour la Protection de l'enfance* to, amongst other things, discourage the custom of breastfeeding by promoting bottle-feeding for infants; eradicate the traditional practice of postpartum sexual abstinence, and promote shorter inter-birth intervals. While this strategy did in fact succeed in addressing the colony's immediate labour shortage problem, it sowed the seed for the population explosion problem, which has been a continuously nagging problem in the Congo Democratic Republic (previously, Zaire) since the mid-1980s.

Development implications of age-grade systems

As stated above, one of the most important social institutions in ancient and pre-colonial Africa was the age-grade system of education in which people were divided into groups according to their ages, with each age-group being taught skills considered relevant and important for coping with the challenges of life at their specific age. With the advent of colonialism and the concomitant destruction of traditional African institutions, the age-grade system became a thing of the past. Yet several problems facing the continent at the moment would have been better dealt with by harnessing the efforts of age-groups. Such a role for age-groups as I am envisaging here is already being assumed by age-grades in Nigeria. In Eastern Nigeria, for instance, Obijiofor (n.d.) noted, that age-grades are cooperating with village leaders in efforts to realize local development projects. Particularly noteworthy in this regard is the amicable competition that takes place amongst the different age-grades as each tries to outdo its friendly rivals in local development activities. Thus, it makes perfect sense to officially adopt the ancient age-grade system as a development strategy throughout the continent. Apart from its potential for accelerating economic development, the age-grade system holds enormous promise for social development. In this respect, the system can serve as a social indicator, which separates one age-grade from the other. Ultimately, this can effectively guard public morality through censorship of group members.

Conclusion

The African family has survived through centuries of brutal assault amazingly in tact. However, it is beginning to show signs of breaking down. The role of exogenous forces in this connection has been overwhelming. These forces, particularly those associated with modernity, have de-emphasized the importance of the family as a social unit in favour of formal institutions. Yet, there is a dire shortage of such institutions, particularly those designed to provide social welfare services. The few existing ones are woefully inept.

As I have argued in this chapter, efforts on the part of Western change agents to supplant the traditional African concept of family have never been based on any rational foundation. Rather, these efforts constitute part of a broader agenda to Westernize Africans. These efforts typically begin with the baseless denigration of African cultural and traditional practices while uncritically extolling the Western substitutes. Thus, supplanting any given traditional African practice with a Euro-centric or Western substitute is sanctioned even in situations wherein the former appears potentially more beneficial to development than the latter. I have already highlighted the advantages of the African concept of family, and especially drawn attention to the fact that it is far more superior to the Western substitute. Those in the international development arena will do well to preserve aspects of the traditional African family that can be potentially beneficial to development efforts on the continent. Thus, an important criterion for preserving an attribute of the traditional family structure should not be the extent to which it mimics the Western equivalent. Rather, it should be the extent to which such an attribute is capable of facilitating development efforts in Africa.

Chapter 5

Traditional Land Tenure Systems

Introduction

Few aspects of Africa's cultural experience are as complex as its land tenure systems. Although for convenience sake, I will be talking of 'the African land tenure system' in the passages that follow, there has, in fact, never been a single system of landholding with application throughout the entire continent. Rather, there have always been a number of landholding systems – some quasi-individualistic and others communalistic, with the latter tending to dominate (for some examples, see Besteman, 1994; Cohen, 1974; Delafosse, 1911; Gildea, 1964; Hamilton, 1920; Hughes, 1962; Shipton, 1994; Shipton and Goheen, 1992; Sorrenson, 1967). People's rights to land in Africa are often overlapping and interlocking, and usually reflect any given society's social fabric. This fabric may be woven around kinship, age-grading or other traditional African principles.

Where does land fall within the broader context of African culture? Accurately responding to this question requires a good command of African traditional and customary practices. One cannot meaningfully separate culture, especially religion and ritual on the one hand, from adaptation, sustenance and production, on the other. To be sure, the cultural importance of land is appreciated the world over. However, it is important to note that the number of dimensions attributed to land as well as the number of people, minerals, plants, animals and buildings it may include differs from one culture to the next, and from one language to another (Shipton, 1994).

Few issues evoke deeper passions and/or account for more bloodshed in Africa, and everywhere else for that matter, than do disagreements about territory. Recent developments in Zimbabwe are illustrative. Here, the land question has dominated politico-economic discourse since the last decade or so. At the heart of the land crises in Zimbabwe are issues whose roots go a lot deeper than the sensationalized accounts in European, North American and other Western media portray. As I stated in the opening chapter, these accounts are uniform in their vilification of the country's president, Robert Mugabe, whose decision to confiscate, and transfer large tracts of previously White-owned and/or controlled land to members of the country's Black native population is said to have been predicated on reasons of political expediency. The failure on the part of Western and pro-Western analysts to acknowledge the vast gulf separating the African and Euro-centric models of land 'ownership' defies explication and eludes comprehension. The difference in traditional African landholding systems alluded to above notwithstanding, a number of common threads are discernible. For one thing, throughout tropical Africa, traditionalists believe that the living have a duty to hold land in a sacred trust for the dead and unborn generations. For another thing, traditional Africans, almost without exception, do not view tenure as implying outright ownership. Rather, they see it as comprising rights and duties of use, transfer and administration; of access, occupation, and reversionary control (O'Flaherty, 1989; Hamilton, 1920; Shipton, 1994; Sorrenson, 1967). More importantly in the context of this book, African land tenure and the general relation between man and earth, including everything thereupon, constitute integral parts of the cultural identity of African societies. Accordingly, we can safely talk of a uniquely African land tenure system. This chapter is dedicated to exploring the intricacies of this system.

Nature of the traditional land tenure system

The term land tenure as employed here should be taken to comprise the system of legal rights and obligations that govern the acquisition, possession, alienation and transfer of land in any given community. It is essentially for this reason that tenure constitutes an important element in the utilization of land throughout the world. The importance of tenure notwithstanding, it counts amongst the least understood socio-economic development phenomena. Nowhere is this more of a truism than in the case of Africa. Here, the story of land tenure has often, intentionally or not, been distorted. In this regard, colonial authorities and other Western agents of change characterized the African land tenure system as 'primitive' with headmen, chiefs and/or whole tribes comprising the landowning units. Yet the truth is further from this.

Land which may comprise no more than soil and sand, (Shipton, 1994), constitutes a lot more than a source of material satisfaction. For some, it is an instrument of wealth, status and socio-economic power. For others, land may be simply a piece of a map, a political power base and/or an aspect of divinity. For traditional Africans, land was important not only because of its ability to sustain life but also as a source of religious power and inspiration. Edmore Mufeme (On-line) essentially lends credence to this assertion in the following words.

> The centrality of land to African economic development has been tied to the significance of land resources to cultural and traditional practices. Rituals related to rain-making, thanksgiving and prayer have historically been tied to the land in Africa. Control of land was thus linked to the complex interplay of economic social and political power.

It is within this context that one can better appreciate the question of access to land in traditional Africa. Once we go past the issue of access, there is the question of how whole communities came to establish claim over land. The importance of addressing this question is accentuated once we appreciate the fact that land in traditional, and particularly, pre-colonial Africa could neither be bought nor sold. Communal claims of entitlement to land under these circumstances were typically established through one or more of the following five routes:

- conquest,
- first settlement,
- cultivation,
- habitual visitation, and
- spiritual sanction.

Some communities, villages, clans and other social groupings in pre-colonial Africa occupied or controlled lands that they amassed as part of the spoils of internecine, intertribal and cognate wars. Hence, the notion of establishing claims of entitlement to land through conquest.

Others established their claims of entitlement or 'ownership' of a given territory by simply asserting themselves as the first inhabitants of that territory. This was commonplace during the pre-colonial period, which was characterized by a lot of population movement. Establishing claims of entitlement to land was also commonly done through sustained cultivation. In this case, it took little more than a member or members of a certain village, clan or community to become the first to convert virgin into agricultural land. Yet others came in possession of land within a certain region by demonstrating that they were the first known group to visit and regularly re-visit that region over a sustained period. Finally, there was the possibility that a group, community, or clan could claim 'ownership' of a certain territory by establishing that they were commanded to do so by a higher being or ancestral spirits to occupy a certain area. None of these possible channels of establishing claims of entitlement to land necessitated economic power.

In contrast, the Euro-centric model that was introduced by colonial authorities throughout Africa is based on the theory that whoever possesses economic and/or political power also has power over land. This philosophy treats land as any other commodity that can be bought and sold on the market. This attribute, that is, the commodification of land, is what most distinguishes the Western from the African view of land. Pre-colonial Africa was replete with societies that treated land as communal property. In other words, social groups such as extended families, clans or communities of ancestrally related people collectively owned land.

Control or administration of land, as Gyasi (1994: 391) noted, was usually 'vested in the leader or his appointee, who may give out portions of the land to the community or non-community members' for variable lengths of time. Thus, nowhere in traditional or pre-colonial Africa was land ever treated as a commodity. Also worth noting is the fact that throughout traditional Africa groups of people or communities as opposed to individuals were recognized as the land 'owning' or 'controlling' units. Worthy of note here is the fact that the social grouping of relevance with respect to rights of access to land is a function of (land) use. Thus, as Shipton (1994) noted, the right of any social grouping to farm may from time to time yield to the rights of others to graze or fetch firewood. Hence, anthropologists talk of 'bundles of rights' and 'nested hierarchies of estate' (Shipton, 1994). For some traditional African societies, the rights or access to land for one purpose may not imply the right to access the land for another purpose.

Colonial and post-colonial governments have been cognizant of the unique nature of the African traditional land tenure system and have spared no opportunity to supplant it with Euro-centric varieties. An important objective of this chapter is to examine the actual steps that colonial officials and the indigenous leadership after them took or have taken to supplant traditional African land tenure systems with Euro-centric varieties. Ultimately, the chapter wrestles with a number of 'what if' questions. For example, 'what if' European colonial authorities had left the traditional African land tenure systems in tact? Here, we are interested in gaining some understanding of the plausible impacts of traditional African land tenure systems – including African beliefs about the value and role of land in people's lives – on contemporary development efforts on the continent?

Efforts to supplant traditional African land tenure systems with European varieties constituted part of a broader scheme on the part of colonial authorities and other Western change agents to transform African societies into facsimiles of Western societies. Officially, these efforts, better known as land reforms, were/ are masqueraded as tools of socio-economic development. Accordingly, agents of modernization and Westernization such as colonial and post-colonial officials, as well as the International Bank for Reconstruction and Development (the World Bank), the International Monetary Fund (IMF) and the United States Agency for International Development (USAID) forcefully argued that traditional African land tenure systems were antithetical to socio-economic development. Conversely, Western land tenure systems, promoted socio-economic development by, amongst other things, reducing the cost of urban land, increasing security of tenure, facilitating transferability of land ownership rights, and enabling the use of land as collateral for bank or other loans (Baron, 1978; Tomosugi, 1980; Feeney, 1982; Feder and Noronha, 1987; Barrows and Roth, 1990). One of my objectives in this chapter is to scrutinize and debunk the foregoing claims relating to so-called modern land tenure systems. Western change agents have habitually exaggerated the weaknesses of traditional land tenure systems while minimizing their strengths.

As I have observed elsewhere (Njoh, 2003a), the notion of ownership in the context of traditional African land tenure effectively overstates what were in fact, custodial or trustee rights of beneficial use over a sacred entity, namely land. To better appreciate this line of thought, one needs to first understand that according to African tradition land belongs to the dead, the living and generations yet to be born. Within the framework of traditional African land tenure, family heads, chiefs and other community leaders are charged with the responsibility of protecting family and/or village/clan land against outsiders. I suspect it is this aspect of the system that has led to the erroneous conclusion that headmen and chiefs owned land in traditional Africa.

In a memorandum he wrote during the heydays of the European colonial era in Africa in 1910, Maurice Delafosse, at the time a member of the colonial civil service in French West Africa, succinctly and accurately characterized the traditional land tenure system of Africa. Although Delafosse (1911) was reporting specifically on the case of West Africa, his characterization holds true for customary land tenure throughout the African continent. As he intimated, one is likely to find more disparities with respect to issues such as marriage and inheritance than in the area of land tenure among various regions of the continent. It is telling that as the colonial official that he was, Delafosse viewed the traditional African land tenure system as 'definitely organized' and 'carrying with it the impress of the genius proper to the negro race' (1911: 260). He went on to draw attention to the striking similarity of land tenure customs throughout the continent. The similarity notwithstanding, it is necessary, especially for purposes of analysis, to classify the landholding customs of the continent into two broad groups, including West and Central Africa, and East Africa and Southern Africa.

West and Central Africa

The landholding customs of West and Central Africa can be further classified into three categories based on the inhabitants of the following geographic regions (Delafosse, 1911):

- forest and coastal zones;
- grassland areas; and
- desert zones.

Forest and coastal zones

An important element in the pre-colonial or traditional landholding system of inhabitants of West Africa's forest belt is the distinction it made between cultivated and 'waste lands.' Within the first category, it was not possible to own land in a sense similar to that associated with so-called modern land tenure systems. Land typically belonged to whole communities and was vested in chiefs or village heads. Anyone directly exploiting the land had no more than the privilege to use the land for farming, grazing or other conceivable purposes. Although ownership of land in cultures of this region during the pre-colonial époque was vested in the chief, it is erroneous to conclude, as colonial authorities were wont to, that the local chiefs owned all land in their areas of jurisdiction. To be sure, land was never considered the individual property of anyone. Rather, it was always a collectively held usufruct. This said, it is worth noting that chiefs or community leaders played a crucial role in the political economy of land in the West and Central African region. Although colonial officials and other Western agents of change were wont to exaggerate this role, it is true that chiefs were endowed with powers to alienate parcels of unexploited, uninhabited or unoccupied land by gift (but not outright sale), grant usufruct rights or the right to exploit land to members of his community or strangers.

The rights of the chief or community leader in this case were terribly constrained, as the consent of the village or community elders was always required. I hasten to note that the rights to occupy or exploit land that were granted to strangers were revocable while similar rights to natives were, as a rule, irrevocable. The elaborate process by which the chief transferred the right to occupy land to strangers in Nso, Cameroon, will serve to illuminate this issue. A more detailed account of this process appears elsewhere (see e.g., Njoh, 2003a; Goheen, 1988). Here, I do no more than summarize the process, which typically begins with a stranger expressing interest in securing land for residential and cognate purposes in Nso (or Nsaw) country to the Fon or the local chief. This is followed by a meticulous appraisal of the request by the Fon and his assistants (the local traditional elders). Depending on the outcome of this appraisal, the stranger is either directed to see a selected native of Nsaw country who is given the task of availing the stranger of land to meet his (the stranger's) residential needs.

In the event that the outcome is negative, the stranger would be shown the way out of the community. Negative outcomes were rare as strangers were generally welcomed in Nsaw country. However, it was made abundantly clear to them that their rights were tantamount to no more than privileges to use the land assigned to them and that these privileges were revocable. The Fon reserved the right to revoke these privileges and could invoke this right when a stranger indulges in conduct deemed in contravention of Nso native laws and/or customs.

As noted above, the right to land or other real property on the part of communities in traditional Africa usually derived from the right of conquest, first occupation of previously unoccupied land, or extended periods of uncontested use. Families were given usufruct rights or the right to exploit arable portions of lands under the jurisdiction of the villages or enclaves of which they were part. However, as stated earlier, they could not own the lands outright although any given family had the power to transfer the right to cultivate part or all lands under its effective occupation.

The sparseness of the African population prior to the European conquest meant that vast portions of land were neither cultivated nor effectively occupied. Who had jurisdiction over such lands? Such land fell under the jurisdiction and control of the local chief. Delafosse (1911) was accurate in observing that traditional African custom allows individuals, families or other collectivities the right to exploit but not to possess land. However, he committed the same fallacy as many a Western observer when he stated that, 'the chief is the true and only owner of the land constituting the domain of his tribe' (Delafosse, 1911: 261). To be sure, no one, not even the chief, owned land in traditional Africa. Chiefs were merely custodians of what was seen to belong to several generations, some dead, others living and many that were yet to be born. Thus, as I have above implied, the chief was endowed with powers to intervene, but only as trustee, in land matters. He had no right to permanently sell or alienate any portion of the land under the dominion of his tribe, clan or village. Families, individuals, or other collectivities (e.g., clans, villages and enclaves) that may be jointly involved in the exploitation of land, did however own the produce that resulted from their own cultivation. Accordingly, these collectivities or individuals were free to do as they chose with their produce or the upshot of their labour in general.

Traditional African customs relating to land distinguished between 'unproductive lands' (e.g., savannahs and rocky areas that were unsuitable for cultivation) and 'productive lands' (e.g. forests, arable areas and auriferous ground that are suitable for cultivation). Lands under the former category were considered the collective or common property of the village, clan or enclave under whose jurisdiction or dominion the lands in question fell. As collective property – as opposed to family or individual property – all members of the village, clan or enclave had the right to exploit the 'unproductive' areas by hunting therein, gathering wild fruits from there. However, such exploitation was limited to activities of a temporary nature and not those that sought to convert certain portions of the lands into possessions that could be transmitted to one's heirs. This custom was widespread in the coastal regions of traditional West Africa, particularly amongst the Algyan, Adyukru and Brinyan of Côte d'Ivoire (Delafosse, 1911).

Apart from 'unproductive lands,' traditional African customary laws of land tenure also classified as public lands areas such as rivers, lakes, the sea, beaches, access roads, market places, grounds used for collective rituals, and the lands directly adjoining village or other communal boundaries. Under traditional African custom, any inhabitant of the village or community under whose jurisdiction they fell had access to all of these public areas. However, certain levels of exploitation of the areas were allowed if and only if they were undertaken by collectivities (e.g., the village as a whole) and sanctioned by the chief. For instance, the hunting of large game or rare animals such as elephants, lions, tigers, and so on, was to be undertaken only at the behest of the chief in consultation with the village elders. Similarly, only fish hunters appointed by, and under the supervision of the Village Council of Elders could harvest rare species of fish from any body of water. The Village Council of Elders was also responsible for distributing the catch to all the villagers.

The grasslands and Sahelian belt

We now pass on to the area of the grasslands stretching from the eastern to the western frontiers of the African continent and extending between the forest zone to the southern extremes of the Sahara Desert. Here, as in the coastal and other regions of Africa, land belongs to all members of the community and is placed under the authority of the chief or village leader. The chief or leader can grant to anyone under his administrative jurisdiction the right to exploit any portion of the land. Again, as in the case of groups in the coastal region, such rights, once granted, were irrevocable except in situations involving strangers or non-natives. Again, it needs to be reiterated that such rights as the chiefs possessed over land did not make them the landlords. However, it is true that African chiefs exercised enormous control over the land constituting the domain of their tribes, clans, or villages. Often, upon acquiring land through conquest or otherwise, a chief, with the assistance of the local elders typically proceeded to divide it up amongst his subjects, particularly the heads of families under his jurisdiction. Each family head was thus assigned a parcel of land and was granted usufruct rights or the right to exploit said parcel.

Historically, people of the Sahel have controlled larger portions of land than the coastal groups. At least two reasons account for this phenomenon. First, the Sahelien population has historically been more sparsely distributed in comparison to the coastal population. The sparseness of the population also accounts for the fact that large portions of land in the region were never effectively occupied, cultivated or otherwise controlled by individuals or families. However, contrary to the claims of colonial authorities, no parcels of land were ever 'ownerless' as every centimeter of land everywhere in Africa was always under the jurisdiction and control of one community or another. Second, the soil in the Sahelien region is wont to get rapidly impoverished. Therefore, it was necessary for people to control large areas as a means of facilitating shifting cultivation.

An important aspect of traditional land tenure in the Sahel was the double conception of the notion of property (Delafosse, 1911). Property was either conceived as the soil and its spontaneous products or all that is the product of human labour. The soil and concomitant natural products were seen as the property of members of the community in which they were located. The chief of the community was considered the manager of said resources. The chief in turn typically subdivided and placed the resources under the control of various family heads. Thus, every family head had the trusteeship and usufruct of a portion of the communal lands. These rights of use as they were given to families should not be misconstrued as rights of ownership.

However, the continued possession of usufruct by any family sooner or later translated into absolute proprietorship of the usufruct, which could be transferred by inheritance, ceded in whole or part by one family head to another, and/or ceded as collateral for a loan. Most notably here is the fact that no one had the right to cut down any tree, especially fruit trees such as 'shear butter trees,' without the consent of the chief in consultation with village elders. Similar to the case of groups in the coastal region, the land tenure of the Sahel allowed for certain easements, which provided for public needs. Such easements applied to the waterways, springs, access roads, market places, and places of religious or ritualistic importance. While the owner of a parcel of land through which an access road or footpath passes may alter the course, he had no right whatsoever to permanently block the road. The right to shoot game with a gun or bow and arrow, and the right to fish with a line or net belonged to all members of the community. However, the right to trap game, organize game-drives and to hunt after setting fire to the bush and to fish by means of dams or traps or by means of herbs belonged exclusively to the family under whose control the parcel of land or body of water in question fell.

Traditional land tenure in Eastern and Southern Africa

Combining eastern and southern Africa as I am doing here is not by any means meant to trivialize the vastness and heterogeneity of the region. Dealing with the area under one umbrella for analytical purposes is meant to simply acknowledge the fact that despite its heterogeneous nature, the region does share some features that make it distinct from the western and northern regions. The traditional land tenure of North Africa has been greatly influenced by Muhammadan Law. It is well established that North Africa has always been influenced by the Middle East and especially, Islam. West Africa on its part is relatively forested and had enjoyed a history of urban living that pre-dated the European Conquest. The Eastern and Southern African region has historically made its mark on the continent as grazing terrain. During the colonial era, one analyst described the region as 'one of the great grazing areas of the world' (Herskovits, 1952: 39). In contrast, the tse tse fly prevented the West African region from gaining a similar reputation.

As in other parts of Africa, land in Eastern and Southern Africa was considered communal property. It had no price in monetary terms, and could not be sold. The native authorities performed the function of caretakers or administrators. Thus, in theory, land belonged to the head of the community or the chief who doubled as the religious and political leader. In practice, control of land resided in the hands of the elders or heads of lineages. Each of these individuals served as a custodian of the land and was required by custom to ensure that each and every member of his/her lineage had a piece of the land apportioned to the lineage. As noted in the case of West and Central Africa above, Europeans misconstrued this aspect of the African land tenure system to mean that traditional leaders such as chiefs and headmen owned all land in the region. Nothing could be further from the truth as all members of the community had equal access and rights to land. There were no individual holdings although individuals could cultivate assigned plots. In this case, individuals had full control over their harvests. The grazing culture, which was dominant in the region meant that individuals cultivated parcels of land on condition that the parcels must revert to pasturage soon after harvests. All land not effectively occupied or under cultivation was set aside for communal grazing.

Farming and herding were thus usually mixed as people commonly indulged in a seasonal oscillation whereby individuals or families claimed relatively exclusive farming rights during the cropping season. The same land upon which crops were grown in one season usually reverted to the community at large or at least to a larger collectivity than members of a single extended family. Accordingly, within the framework of the traditional African land tenure system, one could talk of 'hierarchies of rights and duties between members of a family, household, compound, work group, lineage, or other right-holding unit. These rights and duties tended to shift over the family or life cycle.

Access to land was usually possible through a number of channels the following of which were commonplace (Shipton and Goheen 1992):

- membership and good standing in an extended family, village, clan, network or category;
- labour input or what is often known as 'ability to exploit criterion' (e.g., being the first to cultivate a parcel of land);
- family or community needs (e.g., among the Akamba of Kenya, where land ownership was conferred according to family subsistence needs and the ability to exploit criterion).

The vastness of land throughout the region gave the impression that some land was 'free.' Yet there was never any such thing as 'free land.' Lineages and/or tribal leaders laid claim to all land throughout the region. This is not to say that there was no 'idle land' as there was indeed much of this to go around. However, it is erroneous to view idle land, some of which was deliberately left idle to rejuvenate itself, as free. While land in Southern and Eastern Africa could not be sold or exchanged, its products, including such things as fruit trees (e.g., oil palm and coconut trees) were treated as commodities. Thus, there existed the possibility of privately 'owning' the trees and crops on a given parcel of land, in exclusion of the land itself. The transfer of such transitory proprietorship was commonplace in pre-colonial Africa. However, non-Africans often misconstrued this to entail the transfer of outright ownership.

A case involving the Jibana, a small sub-group of the Wanyika of Kenya, on the one hand and an Arab merchant on the other, which made its way to the colonial High Court at Mombasa, bolsters this assertion (See Hamilton, 1920). As was typical in pre-colonial Africa, the boundaries circumscribing the Jibana and distinguishing them from their neighbours (*Nyika* to the North and South, and Arabs and Mohammadan converts to the East), were marked by distinct natural features such as old baobab trees, hills and valleys. The specific case here considered had to do with the claims of an Arab to the effect that he 'owned' land in Jibana territory. The individual in question was the heir to a Mombasa-based Arab merchant who had interacted with the Jibana for several years prior to his death. In the course of this interaction, mostly for the purpose of trade, he had advanced money to a number of local folks against which they had assigned him as security their coconut trees. Upon the fellow's death, his son, the heir to his estate (or *Wasi*), secured and legalized the documents assigning the coconut trees to his late father. He further proceeded to draw an imaginary line several kilometres long circumscribing the area containing the trees. Then, he went on to sell the land and everything else within the area to an Indian merchant as freehold estate. He claimed that his father had acquired the land from the natives. The Indian merchant, on his part, proceeded to convert the land into a plantation. Initially, the Indian had moved to construct a permanent housing unit, plant special trees marking what he considered to be his property boundaries, enclose the trees and land on which they stood within a fence and forbid the Jibana access to the river along which most of the coconut trees stood. These developments did not bode well with the Jibana natives. Consequently, they took matters into their hands and decided in favour of destroying the fence and attacking the plantation manager.

The Indian owner then applied and secured police protection from the British colonial government in Nairobi. The police was successful in quelling the uprisings. Not long after the decision to arrest the unrests, the colonial government realized that the actions of the natives were founded on grounds worth exploring. Consequently, they decided to deal with the matter in Court. A number of active members of the Jibana community, who had stayed abreast with the developments regarding this matter in the official halls, proceeded to bring forth a representative action for ejection against the Indian. The Indian, it must be stressed, relied on the sale to him by the *Wasi* and the documents of assignment of the trees to the deceased Arab. Not only were these documents drawn up in Arabic, they were largely along the lines of Mohammadan Law, which recognizes the outright sale of land. Thus, what the Arab effectively did was to translate language that would have simply documented the temporary transfer of proprietorship over trees into language conveying the false impression of an outright sale to the presumed purchaser of several hectares of land.

During the course of the trial, the defendant conceded that his extended claim was untenable. What was then left for the Court to try was whether the Arab, a non-Jibana, could acquire and/or had actually acquired a title to any parcel of the tribal land by purchase, or otherwise. Also, the Court had to determine the true nature of, as well as authenticate, the documents purporting to assign the trees to the defendant. The plaintiffs' case against the Arab, as Robert Hamilton (1920) noted was rooted in what he saw as 'vague beliefs.' While the beliefs might have appeared vague to a European such as Hamilton, they not only meant the world to the Jibana, but also possessed pragmatic value to the Court.

A number of factors in the culture of Africa, such as reverence for ancestors and the spiritual power of the earth conspired to give African traditional land tenure a peculiar character. Most parts of societies in traditional Africa had religions that acknowledged or recognized the concept of an 'Earth God' as well as ancestral spirits. The case of the Jibana's is illustrative. The Jibana's harboured a concept of the Supreme Being that is said to exist in many forms, one of which is as a 'sky spirit.' They believed that the sky spirit sends rains to fertilise their mother, the 'Earth.' Every living being is a child of this mother, the Earth. Thus, how dare someone, even in an insane state of mind, imagine selling his own mother! This view of the Earth constitutes the basis of the belief that land belongs to everyone just in the same manner that a mother belongs to all her offspring.

This metaphor can be extended to permit illumination of the fact that members of the Jibana tribe considered it absurd that a non-Jibana could claim entitlement, let alone 'ownership' of Jibana land. Here, it is clear that a woman is either one's biological mother or she is not. There is no question that biological ties generally supersede other ties in life. Accordingly, the Jibana's and other pre-colonial African societies believed that 'strangers' or non-members of any given tribe could not acquire greater rights over land than members of that tribe. This led Hamilton (1920: 16) to conclude as follows.

> And therefore in practice should a stranger desire, for trade or other purposes, to occupy land within the tribal area, he can neither purchase nor rent it, but can occupy land only by permission granted by the 'Wazzee' or Elders, which permission is at pleasure and may be withdrawn at any time.

This method of assigning rights of transitory proprietorship over land bears a striking resemblance to that of the Nso in Cameroon, discussed earlier. Amongst traditional Southern and Eastern African societies, principles and beliefs such as those discussed thus far, safeguarded African land for use by Africans and non-Africans.

The principles that guided the regulation of use flowed freely from the basic idea of common rights. Access to land was typically gained through cultivation. In which case, small parcels of initially virgin land were cleared and cultivated from one season to another until they lost their fertility. Then, they were vacated and left to rejuvenate themselves naturally. Here, as noted earlier, the idea of ownership applied only to the products or harvests from the land and not to the land itself.

Thus, for the Jibana natives they never saw themselves as 'owners" of the land, which they occupied. To the extent that this was established, the Court had no other option but to rule in their favour. The rationale for this ruling was easy to comprehend. The presumed 'purchaser' of the land could not derive claims of 'ownership' from people who saw themselves not as owners but as custodians of the land. As for the defendant's claim of entitlement to the coconut trees, the natives hardly found it challengeable.

This is because their customs applying to agricultural commodities of a more temporary nature differ sharply from those that applied to more permanent crops such as coconut and oil palm trees. Such crops could be mortgaged (quite apart from the land on which they were located) within the framework of the native customs of the Jibana's as well as other traditional African societies. To be sure, traditional Africans wishing to raise a loan were wont to assign some or all of their permanent trees to the lender, who thereupon had the right to harvest and use the trees' fruits for as long as the loan remained unsettled. To the extent that this assertion is accurate, one can infer that this was what the Jibana people had in mind when they permitted the Arab access to the coconut trees. The author of the Arabic document attesting to the transaction either deliberately or inadvertently misrepresented this fact. In the context of Mohammadan Law, the sale of trees invariably includes the land upon which the trees sit (Hamilton, 1920). This was/is certainly not case by African custom.

Implications of non-indigenous land tenure systems

It is no secret that those who have historically sought to dominate Africans, culturally, ideologically, economically and otherwise, including the Arabs who penetrated Africa for purposes of trade and/or permanent settlement during the pre-colonial era, European colonial authorities and Western change agents, have always viewed the African system of land tenure with disdain. However, no group has ever matched the zeal and vigour of European colonialists and Western change agents in efforts to transform this system. Those who have religiously pursued the course of this transformation are wont to contend that the African land tenure system is antithetical to efforts designed to attain so-called modern socio-economic development goals. This belief is so powerful that it has persuaded even some of most ardent defenders of African tradition and culture. In an otherwise well-written paper espousing 'Swazi views on land tenure,' for instance, A.J.B. Hughes (1962: 253) extolls the resilience of the Swazi land tenure system to ferocious abuse by external forces, but falls in the same trap as other Western observers when he proclaims as follows. The African traditional land tenure systems, which he characterized as 'agricultural,' are 'most inefficient and are becoming more so every year. Something must change, and soon, if the land of Africa is to make an adequate contribution to the feeding of its inhabitants.' However, Hughes went on to make the following concession.

> . . . the close connexion in the traditional systems between control over land, political authority, and the whole system of sanctions removes this question from the purely economic sphere. This is the essence of the dilemma. Economic planners and agriculturalists can argue with considerable force that change must come if economic chaos is to be averted. Students of African land tenure, and many of the Africans involved, argue that purely economically oriented legislation to this end would inevitably bring chaos in its wake (Hughes, 1962: 253).

The days when development was taken to be synonymous with economic development, measured in terms of gross national product (GNP) or gross domestic product (GDP) are behind us. Today, development entails a lot more than a rise in economic production; it is above all, a measure of human socio-economic, cultural, political and psychological well-being. Thinking in the development domain has been evolving rapidly while progress in the area of development indicators is yet to catch up. Currently, the best-known indicator of the development construct is the human development index (HDI). Since its development in 1990, the HDI has been used to gauge the well-being of countries throughout the world. Essentially, the HDI measures the progress of countries on the following dimensions: life expectancy (at birth), knowledge (adult literacy rate and gross enrollment ratio), gross domestic product (GDP) per capita at purchasing power parity in US dollars (a proxy for gauging living standard). Although hardly perfect, this definition is far more encompassing than conceptualizing development in terms of economic progress.

Yet, efforts to supplant the African land tenure system with Western varieties have been defended exclusively on economic grounds. Even then, it is not clear that Western land tenure systems are better at promoting economic development than African traditional varieties. Let us interrogate some of the very specific claims that have been advanced to defend efforts designed to supplant African traditional land tenure system, which essentially promote the communal 'ownership' of land, with Western varieties, which advocate the individualization of land ownership and the commodification of land. Individualizing land ownership and commodifying land, proponents contend, facilitates attainment of the following objectives of economic development (Baron, 1978; Barrows and Roth, 1990; Feder and Noronha, Feeney, 1982; 1987; Tomosugi, 1980):

- Reduction in land costs;
- Increased security of tenure;
- Ease of transferability of land ownership rights; and
- Facilitation of the use of land as collateral for loans.

Land cost

According to propaganda campaigns by Western agents of change in Africa, beginning with European colonial authorities, African traditional landholding patterns conspire to ensure the fragmentation of land ownership. This, it is argued, amongst other things tends to contribute to the inefficient use of land as several parcels sometimes go unutilized or are underutilized. Accordingly, some contemporary experts on land markets in Africa such as Doebele (e.g., 1987) have incriminated fragmentation of land ownership as a leading cause of the continent's development problem. This argurment particularly claims that Africa's system of landholding causes a shortage in land supply. I cannot help noting that this charge lacks empirical backing. As I have noted elsewhere (Njoh, 1998), there is no statistical relationship between landholding style and land supply. Witness for instance, the fact that land supply problems are as severe in cities such as Karachi and Delhi, where much of the land is in public hands, as it is in cities such as Bangkok, Manila and Seoul where the norm is private ownership.

Security of tenure

Western change agents argued and continue to argue that the traditional African system of landholding is incapable of guaranteeing land 'owners' security of tenure. This argument constituted a critical element in the foundation of initiatives designed to formalize traditional African landholding systems through Western instruments such as land certificates. Oponents of the traditional African land tenure system are often partial and partisan. They typically deploy a dubious tactic in which they set up a 'straw man' and then demolish it whilst ignoring possible counter-arguments. The inability of traditional African land tenure systems to guarantee security of land ownership is due to the fact that authorities in Africa, adhering to propaganda campaigns designed to supplant African culture with Western varieties, refuse to legitimize these systems. Although pre-colonial Africans did not subscribe to the notion of land 'ownership' and although communities or other human groupings controlled land, land disputes were not any more prevalent in Africa than in Europe where individualized land ownership systems prevailed.

Transferability of ownership rights

Amongst the many charges that proponents of the commodification of land have leveled against the traditional African landholding system is the fact that this system inhibits transferability of land as a factor of economic production. While the concept of outright 'ownership' of land is not recognized in the context of traditional African land tenure, traditional Africans owned and could easily transfer land use rights. Such rights were usually obtained through inheritance or membership in a kinship group. Occasionally, as in the case of the Nso of Cameroon, a community leader such as a chief may assign a new member of the community rights of use over a parcel of land.

Certainly, transferability of land under the traditional African land tenure system differs significantly from what obtains in Western systems. Here, as already noted, land ownership is individualized and land is treated as a commodity. Neoclassical economic theory suggests that the individualization of land ownership invariably leads to the emergence of efficient land market in which land is transferred to those who are most able to put it to the most productive use. Thus, more productive users are able to bid land away from less productive ones (Barrows and Roth, 1990).

The question that we must ask here, is thus: what are the implications for Africa's development aspiration is a system that treats land as a commodity? The most obvious problem that comes to mind is what is commonly known as the 'tragedy of the commons.' This problem attracted the attention of environmentalists in the late-1960s. The problem has to do with the notion that when any scarce resource such as land is placed at the disposal of anyone who can afford it but who has no obligation to preserve it, it will eventually be destroyed. Note that the commodification of land guarantees the highest bidders access to land but does not compel anyone to preserve it for future generations. As I have already stated in this chapter, within the framework of the traditional African land tenure system those who are alive today can only be 'custodians' and never 'owners' of the land they occupy. These individuals have the obligation of preserving the land for future generations. To reiterate, traditional Africans see land as belonging to the dead, the living and future generations. Thus, the African traditional land tenure system is far more capable of avoiding the tragedy of the commons than Western varieties.

Land as collateral for loans

Another charge often leveled against the traditional African land tenure system is that the system does not permit the use of land as a collateral for bank loans. This, opponents contend, is because land 'ownership' under the traditional system is not attested to by written documents such as land titles or certificates, which are portable and can be deposited as collateral for bank loans. This claim is not buttressed by empirical evidence. In fact, there is a plethora of evidence attesting to the fact that although pre-colonial Africans never owned land, and despite the absence of written documents, they were able to use landed property, such as cash crops as collateral for loans. The case of the Jibana people in Kenya discussed earlier is illustrative. To be sure, the using of cash crops such as cola nut, oil palm *(Elaeis guineensis)* and coffee trees as collateral for loans, a process known as 'pawning' continues to be commonplace in rural communities throughout Africa. Critics of the traditional African land tenure system err in many respects and most notably when it comes to their characterization of the nature of this system.

A conspicuous element in this regard is the implication that the system is rigid, stagnant and impervious to change. Yet, there is a preponderance of evidence attesting to the African traditional land tenure system's exceeding flexibility, dynamism and adaptability to changing times. The records of early visitors to Africa reveal that prior to the wholesale commodification of land that took place during the European colonial epoque, a few Africans had recognized the monetary value of land. In fact, there is, albeit limited evidence suggesting that a handful of coastal Africans had willingly participated in land sale schemes – although most had been forced or cajoled into doing so (Njoh, 1998). Such sales, which almost always took place between Europeans, as buyers and as sellers, coastal African chiefs, were recorded in the Gold Coast (present day, Ghana) and Nigeria. Feder and Noronha (1987) noted that one motivating factor in this regard was the production of commercial crops such as oil palm.

In an excellent paper on this subject, Gyasi (1994) marshals further evidence attesting to the extent to which the African traditional land tenure system, with communal ownership as its identifying feature, can adapt to contemporary economic opportunities. Gyasi employs the case of large scale land acquisition for the purpose of oil palm cultivation by groups of common ancestry in Ghana. Oil palm cultivation for commercial purposes is of recent vintage in Ghana like other African countries. In traditional Ghanaian society, oil palm trees grew on their own in the wild although their utility was recognized. Consequently, the oil palm tree was treated as a communal resource whose utility was to be enjoyed freely by all members of any given community wherever it was found. To the extent that land was controlled by all but vested in the traditional ruler (headperson, chief or king), only this traditional ruler had the authority to grant anyone from outside of the community concession to derive utility (e.g. harvest the fruit or palm nuts, or extract wine from the tree while it is standing or has been felled). Such concessions were usually granted on one important condition – namely, that a share of the proceeds be given to the stool (traditional leader) for the purpose of meeting some communal obligation(s).

The system continued to evolve with the passage of time and by the nineteenth century, it had positioned itself to incorporate new elements of the modern economy such as land sales and various forms of tenancy. This change was particularly in response to the increasing demand for agricultural land, especially land for oil palm farming. This rapid change in the economic potency of agricultural land was however, not matched by an ability on the part of individuals to afford such land. One Ghanaian group with common ancestry, the Krobo people were however ingeneous enough to craft a viable strategy to acquire land for commercial purposes. Their strategy, which was deeply rooted in African tradition and based on the communatarian spirit, entailed pooling financial resources through co-operative groups. These groups, known in the Krobo language as *huza*, had as their objective, seeking and purchasing prime agricultural land. The land was then shared amongst all Krobo people in proportion to each one's financial contribution towards the group land purchase. The resulting parcels of land generally took the shape of longitudinal strips or what was known locally as *zugba*. Subsequent to the Krobo land acquisition initiative, other groups in what is now Ghana, and elsewhere in Africa adopted what was by all standards, novel strategies for acquiring land.

The traditional African land tenure system continues to show tremendous resilience despite centuries of assault from external forces. One of the many reasons for this is the fact that despite claims to the contrary, the Western land tenure system's strengths are overrated. Efforts to transplant the system to Africa under the rubric of land reforms have created a litany of problems. For instance, the commodification of land has conspired with bureaucratic forces to increase the concentration of land in the hands of a few. Apart from this, the imposition of alien concepts such as tenancy, possession, secure title, and so on, have constituted more than an emotional assault on the traditional African land tenure system. The traditional African land tenure system is deceitfully simplistic. There are few, if any other, land tenure systems in the world that possess the fairness and flexibility that constitute a major component and strength of the traditional African land tenure system. This fairness and flexibility, as Shipton, 1994) contended, and I concur, guarantees access to land on the basis of need and ability to put it to productive use. Thus, the system ensures that land goes only to those who need it and are capable of maximizing its utility. Consequently, land can only go to those who need, as opposed to those who want, it. Therefore, no one – at least within a group of people sharing a common ancestry should ever be deprived from satisfying his/her need for land.

Chapter 6

Women, Sexuality and Property Inheritance

Introduction

Western analyses of gender relations are unified in incriminating traditional practices and culture as the source of women's plight in Africa. For instance, one notable Western agent of social change, the International Society for Human Rights, West Africa Committee (ISHR-WAC, Online), consider 'traditional practices or customs continue to be the main obstacles in the progress towards gender equality and justice' in Africa. Flowing from these analyses is the all-too-familiar recommendation of 'modernization,' a euphemism for destroying all vestiges of African culture and tradition. This recommendation is founded on the tacit, and quite often, the explicit rationale that 'modern' – read western, values and culture are more capable than African traditional varieties, to eradicate gender-based discrimination and simultaneously elevate the status of women. These claims and assumptions have not been sufficiently challenged. Thus, they have become something akin to gospel truth. Only a few researchers, such as Niara Sudarkasa (e.g., 1986), Fatima Mernissi (e.g., 1988), Bonlanle Awe (e.g., 1991), Philomina Steady (e.g. 1981), and Ester Boserup (e.g., 1970) have dared to challenge this orthodoxy.

This chapter adds one more voice to this cause. Particularly, it challenges the accuracy of the image of traditional practices and culture as the source of discrimination and injustices against women in Africa. Also, it interrogates the theory of Western culture and values as magical nostrums for tackling the task of emancipating and empowering women in Africa. Stated alternatively, the chapter challenges the view that Western culture and values hold more promise for ameliorating the status of African women than the traditional African varieties they seek to replace. In this regard, I marshal evidence to demonstrate that so-called modernization efforts and concomitant strategies have constituted the most potent source of the systematic socio-political and economic injustices that have been visited upon African women since the colonial époque. If this fact has hitherto eluded many, it is because of the general human proclivity towards ignoring history. This in turn leads to other gaffes. These gaffes include not only the characterization of African tradition and culture as static and unchanging, but also the fixation on customs, modes of life and discrete units such as the family, lineage, marriage and kinship. The intention in this case has been to dramatize what is often perceived as the 'bizarre' or 'weird' nature of

African women's social and cultural roles, and their position within the institution of marriage (Lewis, Online).

African culture and the status of women

As intimated above, Western scholarship, in large part, contends that traditional practices are injurious to the status of women in Africa. This scholarship impresses upon us, as Niara Sudarkasa (1986) was quick to observe, that in traditional African societies, women are/were essentially 'jural minors' throughout the entire span of their lives as they are/were required by custom to be, first, under the control of their fathers and then, that of their husbands. The exclusion of women from parenthood here is deliberate and intended to lend credence to the distorted image of men as the dominant sex in traditional African societies. Efforts to distort the image of African women as well as their place and roles in society vis-à-vis their male counterparts must be understood within the broader context of the tendency of colonial anthropologists and cognate professionals to project the Victorian Age society with which they were intimately familiar on Africa. Here, I hasten to draw attention to the rich literature on the role and status of women in Western societies during the Victorian Era (see e.g., Woloch, 2000; Battan, 1999; Perkin, 1993; Vicinus, 1977). Married women in the Victorian era had virtually no legal rights as they were required to relinquish any premarital rights to property and personal wealth they might have enjoyed to their husbands. Furthermore, the Victorian woman was deemed not only legally incompetent and irresponsible, but also entitled to no recourse in any legal matter. The exception was in matters sponsored and/or endorsed by the woman's husband. Thus, wives in Western societies during the Victorian era were literally the chattels of their husbands. To the extent that this is true, one can appreciate how colonial anthropologists whose theory of gender relations was largely informed by the Victorian experience might have been flabbergasted by the visible public role and status of pre-colonial African women.

To be sure, most African societies were patriarchal prior to the European conquest. However, it is a quantum leap to conclude that patriarchy, on its own, created conditions under which the relationship between women and men assumed a hierarchical character. In pre-colonial African society, men and women played complementary roles in every day life. This was more so in the case of childrearing. In fact, children are seen as belonging to whole communities as opposed to individuals. Hence, the old African adage that, 'it takes a village to raise a child.' Also note the use of the term 'control' in this context. This is not accidental, as the history of gender relations in the West has always centered on the issue of 'control.' The tendency has been for men to strive to maintain control and dominance over women.

Thus, as already intimated, when Westerners, including anthropologists and other professionals in the colonial service, embarked on the mission to understand their so-called 'colonial subjects', the tendency was for them to project themselves, including their culture, on those 'subjects'. Appreciating this line of thinking will go a good way in facilitating understanding of how Western anthropologists and cognate professionals interpreted various aspects of African tradition and culture. Westerners find traditional African conceptualizations of, and/or practices relating to, the following aspects of life particularly perturbing and injurious to the status of women.

- The family, wife and property inheritance;
- Socialization and gender-based division of roles;
- Construction of female sexuality;
- Women's sexual freedom;

Do African customs in the afore-mentioned domains contribute to diminishing the status of women as charged by Westerners? How well have the Western varieties introduced to replace African customs fared in terms of elevating the status of women in Africa? I tackle these questions in the passages that follow.

The family, women and property inheritance

The traditional family is frequently identified in the Western mass, scholarly and professional media as a major contributor to the plight of women in Africa. Most of the attack in this connection has been leveled against three interrelated customs, namely polygamy, especially polygyny, which entails one man marrying more than one wife; wife inheritance, the custom dictating that the son, brother, sister or any other designated consanguine relative of a deceased man inherits his widow; and property inheritance, which is said to be inherently biased against women. I examine each of these charges and their real and potential implications for the status of African women in turn.

Polygyny

Polygyny has been lambasted for contributing to dis-empowering women by burdening them with duties associated with reproduction rather than production. In other words, polygyny values only the maternal role of women while neglecting their capabilities in the economic, political and cultural domains. Polygyny has also been attacked as contributing to the economic impoverishment of women. Furthermore, polygyny has been blamed for dehumanizing and limiting women's freedom.

I submit that the attacks on polygyny have had little or nothing to do with the socio-economic development of African women and more to do with supplanting African traditional institutions with Western varieties. Stated alternatively, missionaries and other Western change agents were more interested in promoting the notion of 'traditional family' that was in vogue in their home countries during the Victorian époque. The so-called traditional family comprised a man who worked all-day and a stay-at-home wife and their offspring. This image continues to dominate the mass media, especially television and magazines, in Western societies despite the fact that it does not reflect reality in these societies. Consider the United States for instance. According to the US Census Bureau (Online), only seven per cent of all households in the United States consisted of married couples with children in which only the husband worked. Furthermore, families consisting of a married (heterogeneous) couple with children under eighteen have not been in the majority since 1967 (US Census Bureau, Online).

Before addressing specific Western attacks on African forms of marriage, it is necessary to underscore the fact that until recently, as attested to by the afore-cited statistics, people the world over have always placed a high premium on the institution of marriage. Even in the West where the 'traditional family' model has outlived its utility, married people continue to be more favourably treated in political-economic circles than their unmarried counterparts.

In the United States, for instance, marital status continues to be a leading predictor of one's chances of winning an election or of being appointed to a high-level public office. The following statistics lend credence to my assertion. Of the 538 members of the 109th Congress of the United States, 452 (84%) are married; and of the nation's 100 senators, 90 (90%) are married (Congress.org). Only one of the nation's nine Supreme Court justices is unmarried. Similarly, with the exception of one case, that of James Buchanan, 15th US President: 1857–1861, only married men have held the office of President of the U.S. since that office was established in 1789.

Married people are also accorded preferential treatment in the lower echelons of society. For instance, as part of efforts to reform the welfare system in the U.S., authorities instituted a programme called the Temporary Assistance for Needy Families (TANF). A married woman receives more financial assistance than her unmarried counterpart under this programme. Additionally, as Erika Smiley (Online) noted, married women enjoy further benefits, such as subsidized housing, which are not available to their unmarried counterparts in the same low-income bracket. The preferential treatment married heterogeneous couples receive in this case is obvious.

Despite this overt bias in favour of marriage in Western societies, little or nothing has been done to ensure that every woman desirous of getting married can actually do so. By permitting polygynous marriages, African tradition essentially recognizes an important fact of life, namely that women will always outnumber men, if for no other reason, because women outlive men. Also, it is important to understand that not all men can make 'good' husbands. Finally, not every man is willing and/or able to get married. It is particularly for this reason that polygyny effectively makes sense.

Polygyny, according to its critics, overburdens women with chores in the domestic sphere thereby limiting their participation in the public domain. Implicit in this charge is the claim that women in polygynous marriages have more children than their counterparts in monogamous households. This claim is not bolstered by empirical evidence. Although a few studies (e.g., Pool, 1968; Olusanya, 1971; Chojnacka, 1980) have suggested that there is no marked difference between the fertility of women in polygynous marriages and those in monogamous ones, a preponderance of evidence (see e.g., Dorjan, 1959; Reyna and Bouquet, 1975; Clignet, 1970; Ukaegbu, 1977; Shaikh et al., 1987) suggests that women in polygynous households experience significantly less fertility than those in monogamous marriages. No study to my knowledge, has suggested that women in monogamous marriages experience less fertility than their counterparts in polygynous households. Also, it is reasonable to believe that women in polygynous households have fewer domestic chores than those in monogamous marriages. In fact, the need to reduce domestic burdens is one reason why an African woman may seek an additional wife for the husband.

The claim that polygyny tends to economically impoverish women rings equally hollow. Although often deliberately ignored by opponents of the institution, polygyny is not, and has never been for everybody. Only those who control significant amounts of wealth or resources are wont to marry more than one wife. This is particularly because of the high bride price associated with marriage in most African societies. In traditional Africa, a man could not even think about having a second wife without building a home for the woman as custom dictates that each woman in a polygamous marriage must have an abode to herself albeit in the same compound as the abodes of her husband and co-spouse(s). That some men in urban centres in Africa today insist on having more than one wife under the same roof must be seen a serious deviation from what custom prescribes for polygyny.

Given that each woman in a polygynous marriage is entitled to an abode of her own, it is therefore fastidious to claim that polygyny limits the freedom of women. To be sure, the relevant custom frowns at polygynous husbands who do not respect their wives' private space of which her abode constitutes a major element. Women in polygynous marriages pay conjugal visits to their husbands; however, the husband may not enter any of the wives' abode unless in the event of illness or an emergency.

It is also preposterous to charge that polygyny leads to the impoverishment of women. Anthropologists interested in this matter, or what they refer to in professional jargon as 'resource stress,' are not unified in their conclusions. Some (e.g., Chojnacka, 1980) suggest that polygynous marriages face more resource stress than their monogamous counterparts, while others (e.g., Mulder, 1989) conclude that there is no difference between the resource endowment of the two. I am certainly not without an opinion on this matter. I believe that women in polygynous marriages have access to more resources than their counterparts in monogamous households. My position is bolstered by the following reasons. First, an important prerequisite for taking on an additional wife, in traditional Africa, is that she must be guaranteed not only an abode of her own, but also a parcel of farmland. This requirement tends to be less stringent in the case of monogamous marriages, where the wife shares an abode with her husband and lives off part of the mother-in-law's farm. Thus, a woman in a polygamous marriage can count on having an abode and a piece of farmland to herself – something that cannot be said of one in a monogamous household.

Second, polygynous families tend to be significantly larger than monogamous ones. This must not be confused with the issue of fertility raised above. Here, I am talking of the offspring of a man and his wives as opposed to the offspring of one wife. With many children any one of a man's multiple wives can count on more children (her own and those of her co-spouses) to provide for her in time of need. Also, a woman who might not have had the fortune of bringing forth a child of her own, can count on the children of her co-spouses to help complete the chores required of children in Africa. Such chores include, but are not limited to, fetching water, firewood, and helping on the farm. In traditional Africa, the term mother refers both to one's biological mother and all of her mother's co-spouses as well as her maternal aunts. One's paternal aunts and uncles are referred to as one's fathers. This apparent lack of precision in referring to members of one's kin is what Niara Sudarkasa (1986: 101) views as a linguistic clue to 'the gender 'neutrality' of many African societies.' Ifi Amadiume echoes the same sentiment in *Male Daughters and Female Husbands* where she persuasively argued that sex and gender did not necessarily coincide in pre-colonial African societies (Amadiume, 1987).

Thus, to the African, it is not out of order for a woman to refer to her brother's wife as her 'wife.' Nor is there anything unnatural about a woman referring to, and treating her co-spouse's children as if they were hers. In fact, she is required to do so. Opponents are hasty to employ conflict-ridden polygamous homes, where feud is commonplace amongst the co-spouses and offspring is prevalent, to make their case against polygamy as a viable form of marriage. The inconsistency here is glaringly obvious. There has never been any instance in which dysfunctional monogamous homes – and there are plentiful in especially Western societies – have been used to make a case against monogamy as a preferred form of marriage.

Finally, there are the virulent attacks on traditional African inheritance customs. The brunt of these attacks has been leveled on two specific customs, the one having to do with 'wife inheritance' and the other relating to property inheritance. With respect to 'wife inheritance,' opponents are extremely myopic in that they view the relationship between a widow and her 'inheritor' as exclusively sexual. It is basically this view that has led Western change agents such as Human Rights Watch (e.g., 2003), to link the custom to the spread of HIV/AIDS and calling for its abolition. Yet, in traditional Africa, the concept of husband neither connotes masculinity nor conjugality. Thus, it was therefore not unusual for the sister of a deceased man to inherit his widow. Also, it is not uncommon for a young man to inherit his deceased father's widow(s) who could be the same age or older than his own mother. In these two scenarios, within the framework of African tradition, neither the deceased sister nor son would be expected to have sexual relations with the widow(s). The custom of inheritance in traditional African societies was meant to ensure that women were not neglected upon the passing of their husbands. The inheritor was charged with the responsibility of ensuring the wellbeing of not only the widow but that of her offspring. It is therefore safe to argue that inheritance empowered, rather than disempowered women, particularly widows.

Commentaries by those who find fault with indigenous African culture are also mired in ignorance when it comes to the property inheritance question. Contrary to popular opinion, Women in many parts of pre-colonial Africa could, and did, inherit property. However, it would appear that critics of African culture on this issue confuse the Western notion of 'ownership' with what obtained in Africa during the pre-colonial era. As I stated in Chapter Five, the notion of ownership was absurd and meaningless to the African when it came to real property. For example, land could never be 'owned.' If anything, people 'owned' or more appropriately, controlled the right to use land. Paradoxically, the International Centre for Research on Women (ICRW), a Washington, DC-based entity with interest in the developing world, is partially right when it makes the following assertion.

In most parts of sub-Saharan Africa, women historically have enjoyed access rights to land and related resources through a father, brother or husband, depending on a community's lineage system. But access is not ownership (ICRW, 2005: 4). The ICRW is right in asserting that 'access is not ownership' – that is, if what the ICRW means by 'access' is the right to use land. To be sure, no one, be it a man or woman, owned land in pre-colonial Africa. All a person could possess was the right to use land, that is, usufruct rights. However, the implication by ICRW and other like-minded entities to the effect that women's access to land in pre-colonial Africa was solely through men is misleading. While this claim holds true today, thanks to efforts to supplant indigenous African culture with Western varieties during the colonial and post-colonial époque, it was not the case in the pre-colonial era.

During the pre-colonial era, African women had far more direct control over real property and chattel than they do today. Amongst the cattle-rearing societies of Eastern and southern Africa, the bridegroom's family was required to give as dowry, several cows to the bride's family. These cows were supposed to be the property of the bride's family if she remained married for the rest of her life. However, in the event of a divorce, the cows automatically became her personal property. Among some West African groups such as the Meta and other grassfield groups of Cameroon, the gifts (e.g., goats, and chickens) constituting part of the dowry package on a girl were usually preserved and used as the dowry for her brother's bride. In the event of a divorce, the beneficiary brother is required to ensure the divorced sister's resettlement in her paternal home. Thus, dowries constitute a form of insurance policy for women.

In some cases, particularly the matrilineal customs of Africa, women enjoy better access to real property and chattel than men. Consider the case of the Yao and Nyanja-speaking groups of Malawi reported by Joey Power (1995). The Yao and Nyanja are matrilineal in that they trace their origin to one ancestress. Traditional Yao and Nyanja custom does not require the payment of bridewealth. However, grooms were expected to make presents of small gifts to, as well as complete several chores, most of them farm-related, over a two-year period for the bride's family. Upon marriage, a Yao or Nyanja man was required to move from his own village to join his bride in her own village. This usually made the men outsiders wherever they were permanently settled. Access to land for the men was tied to marriage. In other words, they had rights to land exclusively through their wives. Women were also responsible for making all the managerial decisions relating to farming, the household and the family's property. The autonomy went beyond property to affect the extent to which a Yao or a Nyanja woman controlled her own sexuality. This is particularly because, under the matrilineal system, the notion of an 'illegitimate child' or 'outside child' was biologically absurd and impossible. This means that children conceived out of wedlock or through an adulterous relationship were not viewed as 'illegitimate' as is the case in Western societies. Similarly, the mothers of such children were not shunned or treated as social misfits. In the event of a divorce, the man was dispossessed of any access to land he might have controlled through his wife but had the option of accessing land in his matrilineal home. However, as Power (1995) notes, men were in a precarious position where land was scarce.

Socialization and gender-based division of roles

The Western picture of socialization and gender-based division of roles in traditional Africa is out of focus. While it is true that girls and boys in traditional African societies were/are raised to perform some gender-specific tasks, it is erroneous to state that, boys, and only boys alone, must perform certain tasks while girls, and only girls alone, must perform others. It is equally fallacious to claim that certain tasks are set aside, uniquely for men while others are assigned specifically to women. I will return to this issue later. For now, let us see what it meant to socialize boys differently from girls, and the implications of apportioning roles along gender lines, in traditional African societies.

True, African boys and girls were, in large part, raised to perform different roles, based on their abilities, in traditional African societies. On the one hand, boys typically grew up defending their ancestral lands (in battles), herding livestock, hunting, tapping palm wine, fetching firewood and clearing the farms. Girls on the other hand, grew up cooking, hoeing and tending the farms, babysitting, and fetching water. It is tempting to deduce from this that a traditional African girl's place was in the kitchen. However, to the extent that this is true, it is also safe to conclude that a traditional African boy's place was in the hunting grounds and battlefields. Whether one was in the kitchen or in the battlefield or kitchen, did not matter. What mattered to the traditional African was the fact that the roles performed by the two sexes were equally critical for societal survival and maintenance.

Perhaps more noteworthy is the fact that these roles were neither rigid nor assigned differential levels of importance. In other words, the importance of a task or function was never contingent upon the gender of the person performing the task. Societal roles were, therefore, essentially egalitarian. Thus, unlike Western societies, traditional Africans never valued the role or activities of women in society less than those of their male counterparts. The relative devaluation of female functions in Africa is a product of European colonialism. Prior to that, African women not only worked alongside their male counterparts, but also rose to occupy conspicuous and highly respected positions in society. They functioned as queen mothers, queen-sisters, princesses, chiefs, warriors, supreme monarchs, farmers, traders, technicians, artists, and also occupied high-level administrative offices in towns and villages (Sudarkasa, 1986).

Some (e.g., Sacks, 1974) have argued, that the subordination of women results from their confinement to roles within the domestic sphere as well as their exclusion from the economic realm. This argument renders indefensible claims to the effect that women were assigned roles or tasks that commanded less respect than those of their male counterparts in pre-colonial Africa. For one thing, women and men played overlapping roles in pre-colonial African societies. For another thing, they shared and in some cases switched roles in these societies. For instance, during the pre-colonial era in what is currently western Nigeria, Yoruba women were involved in trading and commercial activities that sometimes took them away from home for extended periods (weeks or months), while the men were farmers, who worked close to home and were therefore responsible for executing most domestic tasks. In traditional Ibo society in the eastern part of the same country, the roles were reversed as the men were the traders while the women were responsible for farming and the day-to-day management of the household.

The colonial era succeeded in significantly altering this situation in particular and gender relations in general throughout Africa. As I have already stated, the roots of the forces that discriminate against women and helps to diminish their status in Africa are located in this era. Three developments were particularly crucial in this regard. The first was the introduction of capitalism as the favoured mode of production. The second was the incorporation of Africa into the global capitalist system. Finally, there was the introduction of cash crops, plantations and mines throughout the continent. These developments conspired to facilitate men's participation in, while excluding women from, financially rewarding activities. For example, the agricultural plantations and mines crafted and enforced hiring policies that overtly discriminated against women. Women were further disadvantaged by policies that encouraged only men to participate in cash crop farming while relegating women to the role of cultivating exclusively food crops. These policies resulted in endowing men with financial means while women remained financially impoverished. To the extent that financial means invariably determined one's status in 'modern societies,' it is safe to conclude that colonial policies resulted in elevating the status of men while diminishing that of women. In other words, colonial policies assigned to men roles that were economically more valuable, while assigning women to those activities that had no direct economic value despite their importance to society's survival and sustenance.

Construction of female sexuality

As stated above, Western anthropologists and other Westerners interested in gender relations in Africa have always been wont to project themselves and their cultures on Africans. This explains why European and other Western researchers viewed most African customs as resulting from the need of men to dominate and control women in Africa. Accordingly, the ancient practice of female circumcision, which continues to date in some African societies, was construed as a strategy on the part of men to control female sexuality. This view is not supported by empirical evidence. In fact, men in traditional African society seldom manifested any traits indicative of a lack of confidence in their 'masculinity.' Similarly, they were rarely concerned with where and how their wives spent time away from home. As mentioned earlier, in some traditional African societies, women were involved in long-distance trading that necessitated their absence from home for extended periods at a time. Why then, did European anthropologists rush to establish a connection between cultural practices and a desire on the part of men to dominate women in Africa? The answer to this question can be found in the social history of the Western world. Before delving into this issue, it is necessary to examine the practice of female circumcision as an element in the construction of female sexuality in Africa.

Female circumcision is often inflammatorily referred to in the Western media as female genital mutilation (FGM). This practice, although no longer common, can still be found in some degree in every country in Africa (Kouba and Muasher, 1985; MRG, 1980; ISHR-WAC, Online). Five different types of operations to the female genitalia constitute what is generally lumped under the rubric of female circumcision. The oft-used term, circumcision may be misleading, as the operation bears no resemblance to male circumcision, which involves simply cutting off the foreskin of the penis. However, while circumcision may fail to capture the gravity of damage women sustain from this operation, the term 'female genital mutilation,' as already stated is rather too inflammatory. I therefore elect to employ the term circumcision here.

The five different types of operation comprising female circumcision are (Kouba and Muasher, 1985):

- mild sunna, which entails using a sharp instrument to prick the prepuce of the clitoris;
- modified sunna, involving the partial or complete excision of the body of the clitoris;
- clitoridectomy/excision, entailing the partial or total removal of the clitoris as well as some part of the labia majora;
- infibulation/pharoanic circumcision, comprising clitoridectomy, the elimination of part or all of the labia minora, the excision of the inner walls of the labia majora, the stitching together of the edges of the vulva; and
- introcision, which entails tearing, manually or with a sharp instrument, the vaginal orifice.

Of these five types of female circumcision, only two, excision and infibulation, are commonplace in societies which practice female circumcision in Africa.

Reasons for female circumcision

For Africans who practice this custom, female circumcision constitutes a testament to group membership. This is especially true where the procedure comprises part of the formalities necessary for initiation into adulthood and/or traditional African secret societies. Some African political leaders such as the late President of Kenya, Jomo Kenyatta, have invoked the initiation function of female circumcision to defend this practice. In this case, it is often argued that abolishing the practice would destroy not only the tribal system but also threaten social, political and cultural cohesion.

For some societies, female circumcision has assumed a place in society that makes an uncircumcised woman appear abnormal. In this instance, a woman cannot be considered a complete adult if she is uncircumcised. In such societies, the clitoris and the labia are viewed as male organs. Thus, eliminating these parts is seen as constituting part of the process of separating the sexes in terms of their future roles in life and marriage. In other words, a girl's femininity is enhanced when these so-called 'male parts' of her body are removed. When the practice constitutes part of the rite of initiation into adulthood, it is always accompanied by explicit teachings about the role of a woman in society.

Female circumcision has also been defended as a strategy for enhancing feminine cleanliness and hygiene. In fact, the local appellation for female circumcision in many of the African societies where it is practiced is synonymous with the term cleansing or purification (e.g., *tahara* in Egypt, *tahur* in Sudan, and *sili-ji* among the Bambarra, in Mali). In such societies, women who have not been circumcised are considered unclean. Such women, in the rare instances in which they exist, are not allowed to handle food and water. Uncircumcised female genitals in societies that practice female circumcision are further viewed as oversized and ugly. The role of superstition tends to be magnified in these societies. For instance, members of such societies generally believe that if not excised, a woman's genitals are likely to grow so long as to hang down between her legs thus, becoming unsightly. Similarly, they believe that a woman's uncircumcised genitals can pose a danger to a man's penis during intercourse and to the unborn child during delivery. Death is believed to be an inevitable consequence in both cases. Another superstition in FC-practicing societies is that FC facilitates fertility. Where this is the case, it is generally believed that uncircumcised women are invariably infertile. Finally, these societies generally believe that circumcising a woman guarantees her safer childbirth.

Controlling women's sexuality by circumcision

Of all the possible rationales for female circumcision, the one that seems to have attracted the attention of Western analysts the most are those having to do with controlling the sexuality of women. Why this fixation on sexuality on the part of Western analysts? I summarize the African argument for female circumcision as a procedure designed to control the sexuality of women before addressing this question. Those who practice female circumcision with a view to attaining this objective believe that circumcision diminishes the sexual desires of women. The ultimate objective here is to reduce a woman's chances of indulging in extra-marital sex. It is also believed that female circumcision prevents pre-marital sex. Furthermore, it is believed that circumcision makes women clean, promotes virginity and chastity and guards young girls from sexual frustration by deadening their sexual appetite. The belief that FC prevents pre-marital and extra-marital sex leads potential grooms in societies that practice FC to steer clear of uncircumcised women. Similarly, men in such societies prefer women who have undergone circumcision as sexual partners. Consequently, uncircumcised women in such societies find it extremely difficult at best and impossible at worst, not only to find husbands but also to find sexual partners in general.

The fixation with female sexuality

With an aggression rooted in romanticism and opportunism, Western change agents, including missionaries, colonialists, and social analysts, have historically been on the offensive against African cultural practices. Those concerned with women and gender relations have generally seen these customs through Western lenses. In other words, they are wont to project themselves and their own culture on the Africans and the African customs they purport to study. To appreciate this fact is to understand some of the many seemingly bizarre interpretations that Westerners were or are wont to make of traditional African customs. Consider the fixation on 'controlling female sexuality' as a rationale for female circumcision.

Recall that most of the initial anthropological studies in Africa were conducted during the European colonial era. The years immediately preceding this era were characterized by efforts to control the sexuality of European and North American women. European and North American men appeared generally uncomfortable with their sexuality and for inexplicable reasons had grown increasingly less trusting of their female partners. Accordingly, these men deemed it necessary to control female sexuality. In England for instance, clitoridectomy was prescribed as a cure for hysteria and 'excessive sexual desires' or 'insatiable sexual appetite' on the part of women. In North America, particularly in the United States Clitoridectomy was practiced for the same reason well into the heydays of European colonialism in Africa.

Evidence attesting to male sexual insecurity leading to efforts to control the sexuality of their female counterparts abounds. These efforts attained their zenith in the United States in the 19th century. During this time, especially in the late-19th century, 'even minor transgressions of the social strictures that defined 'feminine' modesty could be classified' as indicative of an insatiable sexual appetite or nymphomania (Groneman, 1994: 337). Carol Groneman (1994) summarizes the findings of a number of studies that serve to illustrate this tendency.

The first has to do with a young woman who was married to a much older man. The man complained that the woman was increasingly experiencing lascivious dreams and appears to be fantasizing intercourse with other men. The husband recommended that his wife sought medical help from a reputable gynaecologist, Dr. Horatio Storer, who was later to become the President of the American Medical Association. The doctor directed the young woman to temporarily stay away from her husband, go on a vegetarian diet, to abstain from brandy and all other stimulants as a means of diminishing her sexual desires. The woman was further recommended to replace her feather mattress with a featherless one, and to take cold enemas and sponge baths as well as to swab her vagina with borax solution to cool her passions.

The second case involved a seventeen-year-old girl who contacted a doctor because she was having 'a fit.' After examining what he labeled symptoms, such as 'the lascivious leer and lips, the contortions of her mouth and tongue, . . .' and so on, 'he [the doctor] judged her to be in a condition of ungovernable sexual excitement' (Groneman, 1994: 338). As a cure, the doctor prescribed a vegetable diet, cold hipbaths, leeches to the perineum and a number of other drugs. The third case came to pass in 1895 and had to do with a thirty-five year old mother of three. She was diagnosed with '*acquired anaesthesia sexualis episodiac paranoia erotica episodiac,*' which manifested itself as nymphomania.

As these cases suggest, female circumcision is premised on a demented view of women as sexual objects who lack the self-discipline necessary to voluntarily abstain from sexual relations. Consequently, they must be disciplined by excising their sexual organs.

Women's sexual freedom in Africa

Early visitors to Africa, including Arabs and Europeans were surprised, if not perturbed, by gender relations on the continent. The Arab's view of womanhood and a woman's place in society was influenced by Islam, while that of the European was a function both of the Victorian era and orthodox Christian doctrine. Both believed that an ideal woman was supposed to be secluded, protected, pure, maternal, and was expected to belong to the domestic as opposed to the public sphere. Thus, they viewed with disdain, the unrestricted intermingling of boys and girls, women and men that constituted the norm in traditional or pre-colonial Africa.

Women in Africa, unlike their counterparts in Europe and Asia at the time were influential in domestic and public domains alike. Carolyn Martin Shaw's (1995) piece on women in traditional Kikuyu society in Kenya, which draws extensively on Leaky (1977), renders invalid characterizations of traditional African women as docile, passive and powerless. Kikuyu women, Shaw asserts, occupied a range of statuses as professionals such as diviners, prophets, and spiritual healers. Prominent in this connection were 'the councils of elderly women and the district territorial ceremonies organized by women.

Traditional African women had more autonomy and choice than European and Asian women during the pre-colonial era. Also, to the extent that the freedom of women can be measured in terms of the degree to which they have control over their bodies and sexuality, it is safe to say that traditional African women enjoyed more freedom than their Asian and Western counterparts. I surmise Islamic and Christian doctrines on sexual morality are wont to condemn traditional Africans. Here, I would be remiss if I failed to caution against the danger of being persuaded by Western exaggerations and mischaracterizations of this aspect of African culture and tradition.

Traditional Africans were certainly not the savages who lacked sexual morality and indulged in the genre of sexual orgies that are anything but a secret in Western societies. To be sure, there were certain situations and conditions under which married men and women in traditional Africa could legitimately indulge in a sexual relationship with someone other than their spouses.

The following two instances exemplify these situations and conditions. The first has to do with the inability to have offspring on the part of a woman and her husband. In this case, the wife is permitted to have sexual intercourse with a man other than her husband. If the woman conceives under these circumstances, the child belongs to her and her husband and the biological father of the child remains a secret and never introduced into the family picture.

The second has to do with the inability of a man to have a male heir. In such an instance, one of the man's female offspring is made the de-facto heir and charged with the responsibility of bringing forth children who will guarantee continuity of her father's lineage. Such a woman is thus free to have children without having the obligation to marry or reveal the identity of the children's father(s). In instances where the woman inherits substantial wealth, or cannot bear children herself, she is required to 'marry' a young lady who can fulfill the childbearing function. In this case, the young lady is free to have children by a man of her choice, whose identity must also be concealed, especially from her children who are actually considered the children of the deceased father of her 'female husband.' I hasten to note that there is no credible evidence suggesting that 'female or she husbands' and their surrogate wives were indulged in homosexual liaisons.

Those with a propensity to defend African customs cite such instances of legitimate extramarital liaisons to suggest that African women enjoyed freedoms that were denied Western women (Leakey, 1977; Shaw, 1995). This argument, although designed to defend African culture against unwarranted assault from Westerners and other non-Africans, is unfortunate because it misses the objective of the extramarital liaisons described above. To appreciate the objective of these liaisons, one must understand the importance of children and of continuing a lineage to Africans. These relationships have one, and only one purpose, reproduction. The absence of fertility clinics in traditional African societies rationalizes the legitimization of such liaisons.

Virginity and female circumcision

There is a tendency to link the practice of female circumcision (FC) in Africa to a preference for premarital virginity. While premarital virginity commands utmost importance uniformly across the Arab world, the record on Africa is mixed. Some groups in Africa place a high premium on premarital virginity while others do not (Shaw, 1995). For instance, amongst the pre-colonial Coniagui of Côte d'Ivoire, where FC was once prevalent, grooms preferred only brides who had had one or two children. This was necessary for two purposes. First, it was important for a woman to have children for her own lineage before going off to settle with, and have children for, her future husband's lineage. Second, the husband-to-be needed some credible evidence of the fertility of his future wife, and having one or two babies served this purpose. However, it is important to stress that once married, the woman was expected to be faithful to her husband while the man was expected to be faithful to his wife or wives in the case of polygynous marriages. Also, as Carolyn Martin Shaw observed amongst the Maasai of Kenya, who also practiced female circumcision, premarital virginity was not required. In fact, traditional Maasai's believed that girls could not develop breast until they had had sexual relations with young Maasai warriors (Shaw, 1995). Shaw is however not clear on what she means by 'sexual relations.' Some traditional African groups did not construe sexual relations as necessarily encompassing sexual intercourse. While such groups encouraged sexual relations amongst young unmarried persons, they frowned at premarital intercourse.

For example, the Kikuyu, next-door neighbours of the Maasai of Kenya encouraged boys and girls to partake in sensual games, which included everything sexual but copulation. Here, I hasten to add that lessons about these elaborate sexual procedures were passed on from one generation to another throughout most of Africa. Even today, older people, especially in rural areas in Africa are wont to narrate stories of how they, as boys and girls, slept communally without thinking about 'sex'. Shaw notes the disappointment of an old Kikuyu woman with today's boys and girls in the following narrative (Shaw, 1995: 73):

> 'In my day,' she said, 'we shaved our heads, rubbed fat on our bodies, and lay with the warriors in *ngweko*. Today you comb your hair, put on a dress, and get pregnant.'

Shaw goes on to note that (1995: 73),

> The old woman was referring to Kikuyu customs that valued premarital virginity but encouraged unmarried women to engage in formalized sexual activity within a group context. This she contrasted to the mores introduced by the Christian missionaries, which condemned sexual activity among unmarried partners in an effort to preserve virginity and spiritual purity. The old woman found the European traditions laughable: they deny unmarried people legitimate ways to express their sexuality, thus inviting trouble.

The image of sexuality and womanhood as sketched above departs drastically from that which Europeans and Asians held. For Europeans of the 19th and early-20th century, the ideal woman was expected to be a virgin, pure, avoid thoughts that could lead to sexual intercourse. In fact a girl could be labeled a nymphomaniac by simply entertaining 'lascivious' thoughts. More noteworthy here is the fact that during this period, European and Asian women were confined to the domestic sphere of the family and had no roles in the public domain. Thus, these women did not have the same visibility or experienced the same geographic mobility and freedom that their

African counterparts did. However, the introduction of Islam and later, Christianity and colonialism, changed everything.

Although some (e.g., Kouba and Muasher, 1985) have expressed doubts regarding the origins of the practice of female circumcision, all available evidence suggests that it is anything but indigenous to Africa. My guess, based on the geographic distribution and incidence of this practice in certain regions of the continent, is that it originated in Arabia. Note that, as Figure 6.1 shows, female circumcision is most prevalent in the regions dominated by Islam. These include the middle belt immediately above

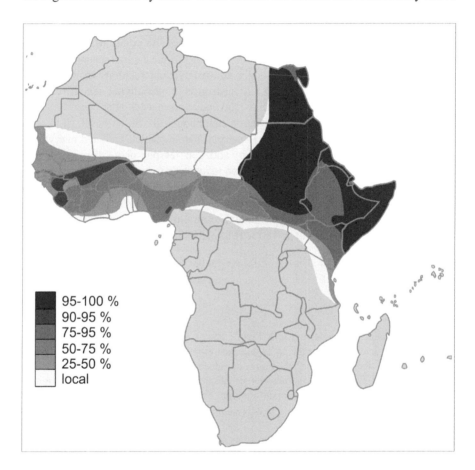

95-100 %
90-95 %
75-95 %
50-75 %
25-50 %
local

**Figure 6.1   Rough estimate of the prevalence of female circumcision
                in Africa**
Source: Afrol News (Online)

and below the Sahara Desert and stretching from Somalia to Senegal and the region along the Indian Ocean on the East Coast of the continent.

Women in the public domain in traditional Africa

International agents of socio-economic change have been accurate in observing that women are terribly underrepresented in the formal public domain in Africa. However, they miss the mark by a wide margin when they characterize this obvious injustice against women as a function of indigenous African tradition and culture.

In fact, as the discussion in this chapter shows, excluding women from the public domain is contrary to the dictates of African custom and tradition. Pre-colonial Africa has a richer history and more profound legacy of having women in very high and visible public offices than other regions of the world. Tables 6.1 and 6.2 show some of the most prominent public offices or positions that were occupied by women in Africa prior to the European colonial era. Many pre-colonial African societies were governed by parallel systems of chieftaincies – the one, female and the other, male.

It must be noted that just as the male prong of such systems did not deal exclusively with male issues, the female one did not concentrate solely on women's affairs. Here, I hasten to draw attention to the fact that in traditional African societies where young people prostrate before their elders, young men are required to prostrate before elderly women as they would before elderly men. In the many cases where there was only one system of chieftaincy, it was possible for men and women alike to ascend to the throne. This means that contrary to popular opinion, inheritance was not the exclusive preserve of men. Women were also known to have inherited power. Inheritance was/is matrilineal in some cases, such as amongst the ancient Akans of present-day Ghana, the Tuaregs of Sahelien region, the Bemba of Central Africa, the Chewa's of Malawi, the Bikom or Kom of Cameroon.

In these matrilineal groups, it is believed that while maternity was incontestable, it was difficult to prove paternity. Thus, matrilineal succession served to prevent 'outsiders' or non-family members from ascending to the throne. The concept of family in matrilineal societies is complex and deserves further explication. In such societies, one's father or spouse is not considered a member of the family while one's mother or maternal uncle is. Although matriliny does not connote matriarchy, the two are highly correlated in practice. That is to say, societies with matrilineal systems tend to be invariably matriarchal. Note that the suffix, *'liny'* refers to descent, while the suffix, *'archy'* refers to power. Thus, it is safe to argue that matrilineal societies tend to confer power on women.

Table 6.1   Pre-colonial African female heads of state

	Head of State	Date	State	Current Location	Remarks
1.	Queen Amina	16th century.	Zauzzau (now, Zaria)	Nigeria	Not clear if she inherited the throne from her mother or father. Strong warrior; reputed as undefeated conqueror.
2.	Queen Nzinga	17th century	Matamba	Angola	Most remembered for building a powerful army, which fought several wars against the Portuguese colonizers.
3.	Queen Ahangbe (a.k.a., Tassin Hangbe)	17th–early 18th century.	Abomey	Benin	She co-ruled with her twin brother, Akaba. Was a very ingenious leader.
4.	Queen Qasa	14th century.	Mali Empire	Mali	Queen Qasa co-ruled the Malian Empire with her husband, Emperor Mansa Suleiman, whom she was accused of plotting to dethrone. Later ruled the empire alone upon husband's death. As daughter of Suleiman's paternal uncle, she was also of royal blood. Only one of a very small number of women whose names have ever been read in the pulpit during Muslim worship in the Mosque.
5.	Candace, Queen of Ethiopia.	1st century A.D.	Ethiopia	Ethiopia	This great queen is mentioned in the Bible (See Acts 8: 26-28). She was one of the sovereign queens the northeast African region, which include Meroe (in modern Sudan). There were at least four other female rulers from the region.
6.	Cleopatra VII	69–30 BC	Kemet (Ancient Egypt, Land of Blacks)	Egypt	Although portrayed as White, Cleopatra was of African decent and reigned prior to the Arab Conquest of North Africa. She ascended to power at the tender age of seventeen. She was the most popular of seven queens to have had this name. She was fluent in several languages and was instumental in making Kemet (Egypt) a leading world power of the time.
7.	Dahia Al-Kahina	7th century.	Berber	Algeria	Queen Dahia Al Kahina was a pure nationalist who fought valiantly to safe Africa for Africans. Her army drove the Arab invaders northward into Tripolitania.
8.	Queen Hatshepsut	1503-1482 B.C.	Kemet	Egypt	One of the ancient queens of the Kemet (ancient Egypt) empire, Hatshepsut was famous for her foresight in recognizing the importance of trade, commerce, national defense and international relations to development..

No.	Name	Date	Region	Country	Description
9.	Queen Makeda of Sheba.	960 B.C.	Sheba	Ethiopia	One of the great virgin queens of Ethiopia, Queen Makeda of Sheba. This specimen of Black beauty is said to have been seduced at one point by King Solomon of Israel. Queen Makeda of Sheba is mentioned in the Bible (see I Kings 10:1-2, 10). She is said to have had a son, Menelik, by King Solomon. Menelik later became King of Ethiopia.
10.	Queen Nandi	1778-1826	Zululand	South Africa	Mother of the great leader Shaka Zulu. Queen Nandi is the evalasting symbol of hard work patience and determination. She withstood and overcame many obstacles to rise to a position of power in all Zululand.
11.	Queen Nefartari	1292-1225 B.C.	Kemet	Egypt	Queen Nefartari co-ruled Kemet, the land of the blacks with her husband, King Rameses II of lower Kemet.
12.	Queen Nefertiti		Kemet	Egypt	Queen Nefertiti was an accomplished political and religious leader. She countered a revolt by the priest and emerged victorious and created a new capital for Kemet called Akhetaten.
13.	Queen Nzingha	17th century	Matamba-land	Angola	Queen Nzingha (a.k.a., Ginga, Jinga), fought gallantly against European slave traders. She was a visionary political and military leader. She strived to free Angola of European influences.
14.	Queen Tiye	1415-1340 BC	Kemet	Egypt	Queen Tiye is arguably one of the most influential queens ever to rule Kemet. She has royal blood as Nubian princess. She married the Kemetan King Amenhotep III who ruled during the New Kingdom Dynasties around 1391BC. Queen Tiye held the title of 'Great Royal Wife' and acted upon it following the end of her husband's reign. It was Tiye who held sway over Kemet during the reign of her three sons.

Source: Compiled from information available at Color Q World, Online at:http://www.colorq.org/Articles/article.aspx?d=2002&x=africanrulers and from Afrocentric Experience, Online at: http://www.swagga.com/index.shtml.

This suggests that in such societies, women tend to be endowed with much wealth and generally exercise a lot of independence and influence. Apart from this obvious form of power, women in traditional African societies that were matrilineal commanded a lot of covert powers.

The following example is illustrative. Men often dominated the chief's court in some traditional African societies. In matrilineal societies, when the court was deadlocked on an important matter, it was customary for the members to take a break. During this recess, members were expected to go back to their respective matrilineal lands and to confer with the ancestors. Effectively therefore, each member was expected to go back and seek advise from older women in the matrilineage.

The level of women's participation in the public domain declined sharply following the introduction of Islam in some parts of Africa. This participation continued to decline in other parts as they experienced contact with other non-Africans originating in male dominated societies. European colonialism further contributed to this problematic until women were completely absent from the formal public sphere throughout the continent. Islamic doctrine discouraged, if not prohibited, women from public life. European colonial authorities, informed by Christian orthodox teachings and the social dynamics of the Victorian era, also actively worked to eliminate women from the formal public sector. This effectively marked the dawn of a new era on a continent, where women and men had always worked alongside each other. Women were thenceforth relegated to chores, such as food crop farming, childrearing, and cooking, that were not financially rewarding. Financially rewarding tasks, such as cash crop farming, plantation work, and mining, to name just a few, were assigned to men. The economic implications of this new division of labour were, and continue to be, far-reaching. While men were able to earn cash in return for their labour, women were not. To the extent that power hinged tightly on one's financial wherewithal, men witnessed their power grow, while women saw theirs wane tremendously.

Change agents will do well to acknowledge this phenomenon, namely European colonialism, as the main cause of women's plight in Africa. Unless this acknowledgement is made, efforts to improve the lot of women on the continent are unlikely to yield any significant positive results.

Conclusion

There is no doubt whatsoever that, African women face a litany of problems, including but not limited to discrimination in critical domains such as the formal job market, politics, transportation, education, healthcare and agriculture. However, it is erroneous to view these problems as a function of African indigenous culture. This is not to say that African indigenous culture is without blemish when it comes to its treatment of women. To be sure, it leaves something to be desired in this regard. For instance, female circumcision and gang rape, which have been known to cause women exceeding grieve and immeasurable psychological damage, have no place in any society that believes in equal rights and justice for all. Traditional leaders and others concerned with improving living conditions in developing regions in general and Africa in particular will do well to work towards altering customary practices such as these. However, it is important to note that most of the attacks on African traditional practices and cultures have been unjustified. African tradition and culture are not inherently biased against women as widely claimed in the literature. Most of the socio-economic and political problems, which African women face, have their roots in European colonial development policies, which were designed to discriminate against women. Colonial authorities initially established schools and other centres of formal education exclusively for boys. It was only later during the colonial era that girls' schools were created. Similarly, colonial authorities enacted policies that actively excluded women from the formal job market. Additionally, women were denied participation in all financially rewarding activities. This meant, amongst other things that women had to depend on men for financial assistance. Thenceforth, women were unable to acquire property such as land or purchase basic necessities without the assistance of men. Thus, one impact of the rush to indiscriminately supplant African traditional cultures has been to make women perpetually dependent on men.

Also, this has succeeded in curtailing freedoms and liberties that contemporary, as opposed to pre-colonial, African women enjoy. To the extent that women's these freedoms and liberties constitute development issues, we contend that some traditional African cultural practices and traditions might have had more promise for Africa's development aspirations than the so-called modern varieties that replaced them.

An important achievement of this chapter is the fact that it has drawn attention to gender as a social construction. It is clear that the concept of gender or its construction is a function of tradition societal norms, beliefs and practices. African tradition does not compartmentalize and/or assign roles based on gender. Hence, it is incapable of using gender as the basis for hierarchically structuring societal roles. Thus, the case for recovering African tradition and culture as a means of reducing, or better yet, eliminating, socio-economic inequalities resulting from the gender-based division of labour in contemporary African societies is compelling. Recovery can be seen in terms of facilitating the improvement in our understanding of the mores and norms of indigenous African culture/tradition as well as the restoration of the philosophies and deep meaning that underlie social practices. Furthermore, recovery can be construed in terms of salvation, particularly saving from obscurity those valuable practices that constitute the hallmarks of African culture. As this chapter strongly suggests, seniority and not gender has typically been the defining category in traditional Africa. Thus, in pre-colonial Africa, people were not precluded from exercising power simply because of their gender, but rather because of their age. This, in essence, is the principal characteristic that distinguishes African tradition from the Western or so-called modern value system that have been imposed on the continent since the colonial era.

Chapter 7

Traditional African Administrative Systems

Introduction

We are a long way from gaining complete and accurate knowledge of Africa's rich history. To proponents of the colonial enterprise, pre-colonial Africa was nothing more than a 'dark continent' inhabited by a bunch 'primitive warring tribes,' who wandered in the jungle without any sense of human organization. Thus, those who are wont to present pre-colonial Africa in this light often go on to extol colonialism for freeing the continent from its tribulations. For instance, a recent article in the Daily Mail (2005), one of Britain's premier newspapers, opined that were it not for British colonialism with the concomitant introduction of a system of laws and government, Africa would have remained a land turgid with endemic intertribal wars and massacres (Dalgleish, 2005).

Fortunately, ideas such as these are increasingly losing their popularity as more objective accounts of Africa's pre-colonial history are surfacing. In this latter regard, there is a growing number of well documented and persuasive works demonstrating that Africans had accomplished a lot in the areas of political and social organization, architecture and city building, arts and crafts, commerce and trade, and so on, prior to the arrival of Europeans on the continent (see e.g., Hanson, 2003; Berman et al., 2004; King, 2001; Young, 2001; Denyer, 1978; Hull, 1976; Fraser, 1968; Davidson, 1959).

Unfortunately within this growing literature, very little has been done to promote understanding of the accomplishments of pre-colonial Africans in the areas of public administration, governance and nationhood. Particularly noteworthy is the paucity or absence of works focalizing on administrative systems in pre-colonial African polities despite the fact that recent scholarship has acknowledged the presence of nations on the continent during that era (see e.g., Hanson, 2003; King, 2001).

Thus, there is a vast lacuna in our understanding of the nature, capabilities and accomplishments of administrative systems that existed prior to the European conquest. This chapter seeks to bridge this lacuna. Particularly, the chapter examines traditional models of nationhood and public administration that existed in Africa before the first Europeans arrived the continent in the 15th century. This examination will therefore include as traditional systems of governance and administration, models that may be of Arabic origin. Associating such models with African tradition is not out of step with the thrust of the present work. After all, as Kwame Gyekye

(1997) reminds us in *Tradition and Modernity*, a model, practice or belief does not have to be indigenous to an area to be considered a tradition of that area. Rather, tradition should be taken to mean any model, practice or belief that has endured through generations. Thus, some of the systems, models or practices examined below, albeit of Arabic origin, had been in Africa for generations prior to the arrival of the first Europeans.

African tradition and the notion of nationhood

Sympathizers of the colonial enterprise are wont to claim that pre-colonial Africa is devoid of any history of human organization worthy of attention. A number of celebrated thinkers of the 19th and 20th century, such as Georg W.F. Hegel (e.g. 1956), Arnold J. Toynbee (e.g., 1934), Peter Bauer (e.g., 1972), to name just a few, discounted pre-colonial Africa's achievements in all human endeavours. To these individuals and others of their ilk, it would be absurd to claim that administrative and political governance systems existed in pre-colonial Africa.

Yet, there exists an inordinate amount of evidence attesting to the fact that not only did pre-colonial Africa possess such systems, they boasted systems that compared favourably with those of Europe and other major regions at the time. Populations in pre-colonial Africa were organized in several ways ranging from loosely connected or fragmented to highly centralized polities. Thus, pre-colonial Africa had not only one, but several systems of public administration and governance structures. The fragmented or highly decentralized systems were in the majority, while the highly centralized polities were few and include the well-known kingdoms, such as the Shonghai Empire, the Ashanti Kingdom, the City-State of Benin, the Bamum Kingdom, in West Africa; the Bakongo Kingdom in Central Africa; the Buganda Kingdom, in East Africa.

A question that has pre-occupied contemporary historians has to do with the extent to which kingdoms, empires and city-states such as those identified here could pass for nations, countries, or states. Addressing this question is necessary as a precursor for making any claim to the effect that pre-colonial Africa possessed administrative structures comparable in any manner, shape or form to contemporary systems of administration. If some (e.g. Reid, 2005; Giddens, 1984) have failed to recognize many pre-colonial African polities as nations, it is because they tend to apply a rather stringent litmus test for determining whether a socio-political entity qualifies as a nation. For instance, Smith (1986) contends that such an entity cannot be considered a nation unless it possesses a unique name, is comprised of persons who share at least, ethnic origin, ancestral roots, culture, history, sense of solidarity, and language. One cannot but ponder the utility of such criteria for determining nationhood status. There is no question that if such a litmus test is used in the non-Western world no polity would qualify as a nation.

Strictly speaking, the term nation, which implies cultural homogeneity, does not have the same meaning as 'country,' which has a geographical connotation, and 'state,' which is a political construct. However, in common parlance, these terms are employed interchangeably. In a technical sense, the terms can be taken to mean different stages of political evolution that begin with what may be loosely called non-national states and culminate in nation-states or sovereign countries. Seen from this perspective, a powerful ethnic or other group may decide to seize control over less powerful groups of similar or totally different ethnic origin. The more powerful group usually proceeds to culturally and linguistically assimilate and incorporate the less powerful groups into a unified system dominated by the more powerful group. This unified system thus assumes an identity akin to what is referred to as a nation-state.

> A nation-state is a specific form of state (a political entity), which exists to provide a sovereign territory for a particular nation (a cultural entity), and which derives its legitimacy from that function' (Wilkipedia, Online).

The notion of sovereignty is worth accentuating here. A sovereign state is one in which most, if not all, citizens are welded by attributes such as language, values, ideology, and belief system. These attributes also give meaning to the term, 'nation.' More importantly, sovereignty usually connotes a unitary state that possesses a unified system of laws and government. Sovereignty further implies that the entity is free from external interference – in other words, no higher authority resides outside of the polity in question.

Seen from this perspective, countless nation-states – ranging from tiny autonomous villages, such as the villages comprising Meta country in Cameroon (Dillon, 1973), to the well-known kingdoms and city-states, pre-dated the arrival of Europeans in Africa. Pre-colonial Africa is certainly not unique with respect to the existence of non-national states. Up until the latter part of the 19th century, Europe comprised mainly non-national states, which were in turn made up of multi-ethnic empires. These empires were typically monarchies under kings or emperors or a Sultan, as in the case of the Ottoman Empire.

Pre-colonial Africa, like other regions of the world at the time, included a variety of administrative systems and government structures. Among these were what anthropologists have called 'fragmented' systems, including groups in which there was no discernible political integration above the local community. Examples of this included the Tonga of Zambia and the Alur of Eastern Africa. Then, there were the 'centralized' systems. The well-known kingdoms fall under this category.

Decentralized and/or stateless polities

A memorable scene in Ali Mazrui's highly acclaimed educational video series, The Africans, shows a large group of people from the Tiv ethnic group in Northeastern Nigeria sitting under a tree and deliberating on a matter of great importance to their entire community (Mazrui, 1986). The deliberations were very orderly despite the large size of the group. The curious viewer cannot but ponder why it takes such a large crowd to deliberate on an important public matter. In contemporary Western society, such a large crowd would be unnecessary as only elected representatives of groups in the crowd will be required to deliberate over the matter. Also, in the West, the meeting would typically be taking place indoors and not outdoors. The stark contrast here is obvious. However, it is difficult to miss the fact that the African set-up is literally more open than the Western set-up. Thus, Mazrui appropriately labels the African situation a great example of 'government by the people for the people.' We must, however, be careful not to confuse administration with governance. The former is a lot more specific than the latter. Administration, and more specifically public administration, can be defined in basic terms to mean the implementation of public policy. It also includes creating and charging government bodies or agencies with the responsibility of formulating laws, adjudicating civil and criminal legal cases, providing for public safety, national defense, and so on.

Jay M. Shafritz and E.W. Russell (1996) provide a basic textbook definition of the term. They define public administration as, amongst other things (Shafritz and Russell: 1996: 16):

- The king's largesse;
- Law in action;
- Regulation of public behaviour;
- Implementation of the public interest; and
- What government does.

Here, it is necessary to draw attention to the fact that public administration was initially construed as 'the King's largesse' (Shafritz and Russell, 1996: 16). From this vantage point, public administration can be taken to comprise the goods, services and/or honours that the king, chief, or clan head bestowed on those under her/his jurisdiction from time to time. In pre-colonial Africa, the responsibility of ensuring that every member of the community had access to communally-owned factors of production, such as arable land, rivers and lakes, rested in the hands of these authorities. However, pre-colonial African leaders did more than simply ensure access to these valuable resources; they also controlled the timing and rate of use of the resources (e.g., which portion of farmland had to be left fallow and when).

This regulatory function of the pre-colonial leadership fits perfectly under what Shafritz and Russell (1996) label, the legal definition of public administration. Yet, another activity of pre-colonial African leaders that can be neatly placed under the category of the legal definition of public administration is their pre-occupation with discipline and meting out punishment to transgressors of the rules or laws of the community. Seen from this perspective, public administration has been defined as 'law in action.' The notion of village democracy was alive, well and strong during the pre-colonial era. In this regard, members of the community or their delegated representatives were known to sit in public squares and to not only deliberate over issues of general interest but also to decide on necessary actions. This fits yet another definition of public administration, namely 'implementing the public interest.'

Finally, public administration can be taken to encompass everything that the king, clan head or village leader and her/his official deputies or assistants (e.g., council of elders, or *nchindas,* as they are known in Cameroon's North West Province) do in their official capacity. In contemporary terms, as Shafritz and Russell (1996: 6) state, 'public administration is what government does.'

The administrative system of pre-colonial African polities typically comprised three major bodies, namely (Jarrett, 1996: 61):

- Council of Elders;
- Chief Priests; and
- Moral Elders and Chiefs.

These bodies were tightly woven to form a single administrative system capable of executing functions ranging from mundane tasks, such as using the talking drum to summon a meeting of the king's aides, to complex undertakings, such as planning and executing a war. Each body was placed in charge of a well-defined set of activities. The Council of Elders, for instance was charged with the responsibility of conceiving, planning, implementing and managing the community's development projects. Thus, projects such as public infrastructure building and maintenance, building and maintaining the chief's palace, and building and maintaining weekly markets fell under the jurisdiction of the Council of Elders.

The Chief Priests in pre-colonial Africa comprised people who were endowed with special, usually spiritual powers and/or skills, which were more often inherited than learnt. They functioned as religious leaders/authorities and acted as a bridge between members of the community and their ancestors. Additionally, members of this body were charged with the responsibility of educating the community on African spiritual laws, religious doctrines and principles. In this regard, they provided knowledge on the importance of living in harmony with the natural environment, including land, rivers, and forests, and why members of the community must see themselves simply as custodians and not owners of these natural resources. Furthermore, the Chief Priests also served as health officials or medical practitioners as they were responsible for healing the sick.

The Moral Elders were responsible for teaching moral conduct and upholding moral standards throughout the community. Moral Elders played yet another important role. They were responsible for recording all major events that took place in the community. In this case, they served as community historians. The fact that most pre-colonial African societies did not boasts a written culture does not mean that they were incapable of recording information. This feat was accomplished through two main strategies. The first and most common one was through story telling. These stories were then passed on from one generation to the next until they became legendary. The other was through drawings or sketches. Such drawings and sketches have surfaced in caves and other artifacts uncovered through archaeological and other discoveries.

The laws, more commonly known as customs, of pre-colonial African societies were, in most cases never written. Yet, their elaborate nature can hardly be challenged. Colonial authorities were mesmerized by the precise and elaborate nature of these laws. In fact, the British, under the indirect rule strategy, adopted large portions of the traditional African laws – what they called Native Laws – as part of their colonial administrative laws. Writing back in 1901, J.G.B. Stopford, drew attention to the importance of native laws in colonial court decisions in cases dealing with inheritance and land use/tenure in Nigeria.

On the inheritance question, one cannot help noticing the extent to which these laws were consistent across the continent. The law on descent of land in Yoruba country is consistent with those of other patrilineal cultures in pre-colonial Africa. Thus, the eldest son had the right to succeed to the land of his father 'for his own benefit and for that of his family. The children and wives have the right to live in the house. The widow who takes another husband loses her right' (Stopford, 1901: 85-86). The law goes on to say that (p. 86),

> Females cannot inherit land, they can only have the right to stay in the house. . . . A female has no right to bring her husband to live in the family house, but she does not lose her right to return to it by marriage. Her children have the right. Money expended on the family house is considered to be expended for the benefit of the whole family and cannot be recovered.

The law of inheritance in pre-colonial Ogoni country, also in Nigeria, was quite different from that of the Yorubas. Among the Ogoni, inheritance went to the first-born daughter. Daughters inherited not only chattels such as household property but also real property such as buildings, cash crops (e.g., palm trees, raffia palms, coconut trees). 'The sons inherited nothing directly from their blood mothers and nothing from their blood fathers' (Kpone-Tonwe, 2001: 390).

The traditional African law on property ownership or control was also very elaborate and appears in large part, uniform throughout the continent. In this regard, a common thread running through this law in all traditional African societies is the fact that land could not be sold. Also, as demonstrated by a case of August 15, 1892 in what at the time was the Crown Colony of Lagos in present-day Nigeria, a grant of land implies neither transfer of ownership nor access to cash crops, particularly palm trees. This was especially so when the transferor was a 'stranger,' that is someone not indigenous to the area. As Stopford (1901: 90) notes, 'the stranger to whom land is granted at Ewu has no right to reap the Palm nuts at all.' Stopford (1901: 91) quoted one of the indigenous persons, party to a lawsuit that was brought before the colonial government in Lagos Colony as stating thus:

> When land is granted a stranger, it is only the land that is given to him to cultivate and to plant yams and cassava and so on: the Palm trees are reserved by our customs to the people of the country, the inhabitants of the district and neighbourhoods who arrange among themselves what part to reap.

Governance in the centralized polities

We now turn our focus to the highly centralized polities, particularly the kingdoms, empires and city-states in traditional Africa. It is true that precolonial Africa's accomplishments have always been discounted. However, this is especially the case when it comes to its achievements in the areas of statescraft, public administration and economic development. In fact, popular opinion holds that states were non-existent in Africa prior to the European colonial era. However, as Jeffrey Herbst (2000: 37) was quick to point out,

> Assuming that states and systems of states did not exist in Africa because the European model was not followed demonstrates, at the minimum, a lack of imagination and, more importantly, a rather narrow conception of how power can be organized.

Questions about the ability of pre-colonial Africans to have had states are often rooted in the fact that firm territorial boundaries were absent on the continent prior to the Berlin Conference of 1884/5, which partitioned the continent amongst European powers. Again, Herbst responded to such questions by highlighting the absurdity of equating states with firm territorial control. He went on to remind us that precolonial Africa was much like medieval Europe, where it was common for the church to share sovereignty over any given territory with various political units. Thus, in Europe, as Herbst (2000: 37) forcefully stated,

> Hard territorial boundaries were also a rather late development. Similarly, citizenship laws – regulations that tie individuals to a unique geographic entity – were generally not codified in Europe until the nineteenth century.

Evidence to the contrary notwithstanding, popular opinion continues to hold sway as it is generally believed that statehood as a strategy of political organization was introduced on the continent for the first time by European colonial authorities. From this vantage point it follows that national government functions such as public administration, national economic development, national defense, and international relations were unknown on the continent prior to the European colonial époque. However, as argued above and supported below, this view is gravely erroneous. There is an abundance of evidence showing that Africa had developed states with centralized governments, which exercised political authority through hierarchically structured units of government – or what in contemporary terms would go under the appellation, bureaucracy, prior to the European conquest. Additionally, these states boasted forces that defended their national territories as well as the ability to integrate vast territories into unified economic systems.

James Giblin's (2005) cogent essay on 'diffusion and other problems in the history of African states' discusses the national organizational and administrative prowess of a number of ancient African kingdoms, including the kingdoms and city-states of present-day south-western Nigeria (Oyo, Ife, and Benin), Asante (or Ashanti) in present-day Ghana, and Luba in present-day Congo (DR), prior to the

European conquest. The factors that led to the birth, growth and development of states in Africa are not different from those that were at the root of this phenomenon elsewhere. States were likely to emerge in regions that boasted an abundance of fertile soil, heavy rainfall, a good and reliable supply of water (e.g., rivers and lakes), and other natural resources such as minerals, and commercial opportunities (afforded by, for instance the region's location).

The Yoruba Kingdoms

Two important kingdoms, Oyo and Ife, and one city-state, Benin rose to prominence in southwestern Nigeria long before the European conquest. Three factors, namely agriculture, art and commerce propelled the Yoruba kingdoms to prominence. Iron, which is said to have been in use in the region as far back as two thousand years ago, played an important role in the development of agriculture in Yoruba country. Early Yoruba farmers cultivated yams and cocoyams, which are indigenous to West Africa, as well as plantains and bananas, which were introduced in the area from Malaysia through East Africa. Yorubaland's resourcefulness and especially its agricultural productivity, had an important effect on the surrounding areas as it effectively transformed formerly hunting and gathering communities into villages, over which Yoruba rulers extended their dominion. With the passage of time, the rulers were able to consolidate these villages into states.

At the heart of this ancient civilization was the City of Ife, which exists to date. Ife was known far beyond Yorubaland for its famous bronze castings and sculptures. Some of the exquisite sculptures from Ife included brass and copper, which was obviously imported from other regions. The Yoruba's exported bronze carvings, agricultural products such as yams, cocoyams, kola nuts, palm oil, fish and cloth to other regions. Thus, trade was an important factor in the region's political development. The rise of the Kingdom of Oyo, to the far north of Yorubaland is credited to regional and international trade. The settlement, which was later to become Oyo Kingdom in the late 14th or 15th century, emerged about 1100 A.D. As a kingdom, it acquired horses, probably from the Sahel to the north and with these, its rulers were able to conquer and dominate much of the surrounding region, including the entire Yorubaland in the 17th century. Between 1730 and 1748, they had extended their reach far west of Yorubaland to incorporate the powerful state of Dahomey as a tributary. Concomitantly, they seized control of the seacoast between Wydah or Ouidah and Badagry. With control of the seacoast, Oyo Kingdom was able to freely trade with Europeans – initially in spices, and later, in slaves.

To the west of Yorubaland and to the southwest of Oyo was another powerful kingdom, the Kingdom or City-state of Benin. Evidence of Benin's fame exists to date in many forms, particularly the ruins of an extraordinarily complex system of walls circumscribing the city. The wall attains a height of nine meters at some points and stretches for a distance of more than 16,000 kilometres (longer than the Great Wall of China). Benin kingdom dates back to between the 12th and 13th century and grew to amalgamate previously quarrelsome chiefdoms into a unified polity.

During the 15th century, the king embarked on a number of institutional reform measures designed to reduce the influence of the hereditary chiefs who held roles subordinate to the king or Oba but participated in his selection. Prominent amongst the reform measures undertaken, was the institution of primogeniture, that is, the rule that only a son can succeed his father in a position of traditional authority. Another prominent reform was the creation of a category of chiefs known as 'palace chiefs' and another category called 'town chiefs.' The king personally appointed these chiefs. These reform measures were in essence designed to consolidate power and buttress the king's control over his subjects. The king further aggrandized his powers by significantly expanding the territorial boundaries of Benin.

Another impact of the king's reforms was that it effectively created a system of checks and balances. In this regard, it permitted the Oba or King to divide and conquer the chiefs under his jurisdiction. He established ingenious means of raising revenues or resources necessary for managing the kingdom's affairs. One of these means entailed requiring his subordinate chiefs to pay tribute, twice per year in palm oil, yams and other foodstuffs. Another revenue generation strategy that the king employed was to establish a royal monopoly over trade in pepper, ivory and other goods with Europeans.

The Ashanti Kingdom

The Kingdom of Ashanti or Asante country rose to prominence in the Gold Coast (roughly speaking, present-day Ghana) between the 17th and 19th century. Although it is arguably the best-known, it was certainly not the only kingdom to have existed in the region. Several factors, foremost of which was the abundance of gold, contributed to the emergence and development of states in that particular region. Traders from surrounding regions, and later on overseas, came to the area in search of gold. The demand for gold created the impetus for the emergence of a vibrant local commercial sector and business people. The local business people acquired enormous quantities of gold, which they in turn used in purchasing slaves from Africans in neighbouring regions. Some of the slaves were used locally while a greater number was sold to slave traders at El-Mina for onward shipment to the Americas and the Caribbean.

The king and his lieutenants manifested their economic prowess by crafting and executing plans designed to improve efficiency in gold production in particular, and the productivity of the national economy as a whole. In this regard, they developed tools that facilitated deep-level mining and maximized the utility of slave labour. The slaves were put to work clearing and preparing the virgin forests, which occupied most of the southern portion of present-day Ghana. These developments led to a concentration as opposed to a dispersion of the local population. More importantly, as Giblin (online) notes, 'these developments set the stage for state building.' In practice, this process was effectuated through the peaceful, and occasionally, the forceful incorporation of neighbouring villages and communities. Through the 'swallowing up' of neighbouring villages, the Asante Kingdom was able to aggrandize its size while correspondingly reducing the number of independent communities in the area.

According to some sources (e.g., Giblin, Online), by the 1700s, Akan country had only a handful of independent communities in comparison to the mid-1600s when it boasted as many as 38 such communities. Over time the successive leadership of Asante consolidated their powers and increasingly extended their reach over neighbouring areas. During its heydays, the Asante Kingdom's jurisdiction extended to as much as an eighty-kilometre radius from its seat in Kumasi. One of the most notable of the Asante kings, King Opoku Ware considered the independence and sovereignty of the kingdom extremely important.

Accordingly, he proceeded to ensure that it was free from external interference and as much as possible, self-sufficient. In this connection, the king moved speedily to reduce the kingdom's dependence on European imports such as cloth, alcoholic beverages and spirits, by creating local weaving industries, breweries and distilling plants. Ware also had territorial ambitions and worked indefatigably to expand the geographical area under his control. By the time of his death in 1750, the Asante Kingdom encompassed a territory that measured more than a hundred thousand square kilometers, and boasted a population of approximately two to three million. To put this in perspective, we must note that a good number of independent countries in the contemporary world contain less than three million persons.

The Asante leadership exhibited a lot of ingenuity in the area of governance as well. Evidence of this resides in the complex bureaucratic structure that they had developed to administer or manage the affairs of the kingdom. At the top of this administrative structure was the Paramount King known as the *Asantahene*. Immediately below the *Asantahene*, who inherited his position along matrilineal lines, were numerous kings or chiefs. These sub-kings had jurisdiction over administrative divisions and acted on the *Asantahene's* behalf. There were two main such divisions, namely the 'Metropolitan District,' encompassing all communities and towns within a fifty-mile radius of Kumasi, and the 'Provinces,' including the towns and villages beyond the fifty-mile radius. The chiefs or heads of the towns within the 'Metropolitan District' enjoyed considerable autonomy, were automatically members of the King's Advisory Council and participated in the enthronement of the King. Heads and chiefs of communities in the Provinces did not enjoy these privileges but were required to pay tribute to the Asante rulers. To boost its revenue generation ability, the kingdom set up an elaborate taxation system. All transactions in slaves, gold, salt, spices, ivory, and all commodities taking place within, or cargoes of these commodities passing through, the kingdom were taxed. The planners of this kingdom displayed their economic planning acumen when they were able to prevent an economic collapse by increasing the sale and/exportation of gold, kola nuts, cloth and other local products when the slave trade was abolished in the 19th century.

The Asante kings were also concerned with threats to their position or authority. Accordingly they strived to consolidate their powers, especially through the centralization of authority. Power in the Asante Kingdom was concentrated in the hands of the King or the *Asantehene*. The Asante kings were not oblivious to the possibility of military coups. In this connection, they sought to significantly curb the military's power by employing two main strategies. The first entailed creating a palace guard and the second entailed hand-picking the commanders of this unit.

The Asante rulers were also well-versed with the planning and implementation of capital improvement projects. Additionally, as some (e.g., Herbst, 2000) have noted, they were also well aware of the use of public projects as a tool for broadcasting authority. By the time of the European conquest, the Asante rulers had constructed an elaborate system of roads that converged on the kingdom's capital city, Kumasi. Jeffrey Herbst argues that 'these roads were vital to the exercise of formal authority because they allowed for the quick movement of troops and bound the territory into a relatively coherent economy' (pp. 41–42). The military role of infrastructure projects such as roads notwithstanding, we cannot lose sight of the fact that these projects were undertaken first and foremost as a means of fostering socio-economic development. Above all, the roads stood as a testament to the king's ability to judiciously utilize the 'people's resources.' Seen from a slightly different perspective, such projects can be considered literally, part of the king's largesse,' one of the many definitions of public administration (see e.g., Shafritz and Russell, 1996).

The Luba Empire

The Luba Empire was located in the area occupied by present-day Democratic Republic of Congo. The factors that contributed to the emergence, growth and development of this ancient kingdom are similar to those that played a crucial role in the kingdoms afore-discussed. However, a number of factors set this kingdom apart. First, in contrast to the West African kingdoms and city-states, Luba was located in the hinterland. However, it had uninhibited access to a large body of water, the River Zaire. The advantages afforded by the river, however appeared to have been nullified by the problems that the natural environment presented. This empire rose to prominence despite the fragility of the natural environment it inhabited. Not only was the land marshy, it contained several lakes and river channels. Accordingly, the inhabitants required dikes to protect homes and other property against seasonal flooding; they also needed drainage channels and dams to retain lake waters for dry-season fishing (Giblin, Online). As the old adage goes, 'necessity is the mother of invention.' Thus, the need to survive in a rather demanding and seemingly treacherous environment compelled the inhabitants of Luba to become not only more creative, but also cooperative. The cooperation, hence political unity was necessary for the large-scale public works projects that were needed for their survival.

The Upemba people, members of the Luba Empire, displayed their remarkable ingenuity by making the best of a rather stubborn environment. They maximized the utility of the river and lakes over which they had jurisdiction by fishing, preserving the fish by drying, and exporting the dried fish to their neighbours further hinterland. They had already become major exporters of fish by the 6th century and by the 10th century, they had diversified their economy to include as part of their export commodities, metal products and salt. Additionally, they had improved their agricultural practices to the point where they had surpluses to export to other regions. Local blacksmith bought charcoal and copper, which they used for the purpose of smelting and working the metal products they had become known for. People of Upemba imported locally unavailable products such as glass beads and cowry shells that originated in locales as far away as across the Indian Ocean. Thus, the Luba state had become extremely vibrant by any measure, thanks to its resourcefulness, ingenuity and propensity for commerce and trade. These factors accounted for its ability to have successfully consolidated power and taken control over its neighbours.

The Songhai or Songhay Empire

The Songhai Empire was vast by all measures and covered the area occupied today by the post-colonial nation-states of Mauritania, Senegal, Gambia, and Guinea Bissau, to the west, through the hinterland regions of Mali, Burkina Faso, Niger, and the northern portion of Nigeria, as well as portions of countries on the Atlantic Coast, such as Guinea, Côte d'Ivoire, Sierra Leone, and Benin.

The Empire had what would pass for a highly sophisticated national governance machinery even by contemporary standards. David Dalgleish (2005) has done a fine job describing this machinery in both absolute and relative terms. Amongst the most noticeable aspects of the machinery is its striking resemblance to contemporary bureaucracies. For instance, the different components or units comprising the machinery were organized by function and geographic area as administrative scientists, working with knowledge gained from Max Weber and like-minded thinkers, were later to recommend for modern governments several centuries later. Songhay's governance apparatus comprised amongst other units, several ministerial bodies. Prominent in this regard was a ministerial body in charge of agriculture and headed by an inspector of agriculture *(Tari-Mundio)*; another in charge of etiquette and headed by a Chief of Etiquette and Protocol *(Barei-koi);* and yet another in charge of the Calvary, which fell under the leadership of a Chief of Calvary *(Tara-farma)*. Another ministerial body worthy of note, especially given the period under discussion, is a ministry that might have gone under the name Ministry of Minority Affairs. This ministerial body had several agencies, which were responsible for the various minority groups resident in the empire at the time. For instance, one such agency was in charge of Berabic Arab Affairs *(berbuchi-mundio);* and another was responsible for White minorities *(korei-farma)*.

Jurisprudence, particularly justice, was an important issue in the Songhai Empire. The leadership acknowledged this importance in many ways, particularly by creating positions of Chief Judges or Cadi (Qadi). As this Arabic appellation suggests, Arab immigrants significantly influenced jurisprudence in the Songhai Empire. However, it will be erroneous to deduce from this that indigenous African systems of justice were non-existent. The following description of how King Takruri of Ancient Ghana ensured the administration of justice by Levtzion and Spaulding (2003: 32–33, quoted in Dalgleish, 2005: 62) is telling.

One of his practices in keeping close to the people and upholding justice among them is that he has a corps of army commanders who come on horseback to his palace every morning . . . When all the commanders have assembled, the king mounts his horse and rides at their head through the lanes of the town and around it. Anyone who has suffered injustice or misfortune confronts him, and stays there until the wrong is remedied . . . His riding, twice every day, is a well-known practice and this is what is famous about his justice.

The justices of Songhay functioned on a two-track system, the one was of the king and the other was of the Cadi. The Cadi's, who were posted to major cities such as Djenné and Timbuctu throughout the Empire, were appointees of the king. He was responsible for dealing with common-law issues, misdemeanours or disputes between citizens and foreigners or amongst citizens. The King or Royal Justices were in charge of more serious crimes such as treason.

Individuals found guilty were sentenced based on the severity of their crimes. Some were sent to prison, while others were required to participate in community services. Serious offenses were usually punishable by what would be viewed as cruel and unusual punishment today. Examples of such punishment included burying the guilty party alive, chopping off one arm – a practice, which continues to date in Muslim societies where the Shar'ia law is in vogue. In some pre-colonial African societies, adultery was punishable by forcing a sharpened bamboo stick into the man's penis and then breaking the stick up. In others, such as the Ngwa Ibo of Nigeria, the same crime was punishable either by death or sale of the offending party or parties into slavery (Nwabughuogu, 1984).

Again, in Muslim societies where Shar'ia law was in force, the woman was shot to death. To put things in perspective, it must be noted that instances of similar cruel and unusual punishment were commonplace in Europe around the same époque. During the reign of Henry VIII in England, for instance, the penalty for not attending church was the loss of one or both ears. Once King Henry VIII's son came to power, he made the crime of brawling in a church or church premises punishable by mutilation.

The Kingdom of Buganda

The ancient Kingdom of Buganda is located on the northern end Lake Victoria-Nyassa in present-day Republic of Uganda. By the time European colonial authorities arrived Africa, Buganda was already a well-established nation, complete with a well-defined territory of its own, a central authority, a common and distinct language, a myth of common ancestry, and a history that went a number of generations back. Colonial authorities and other Europeans before them were marveled by Buganda's sophisticated structure and history, which bore a striking resemblance to those of medieval Europe, especially medieval England. This comparison is, however, only accurate to the extent that we are talking of the proclivity towards conquering and assimilating other groups – something that the Kingdom of Buganda, like medieval England did with unmatched gusto. However, the comparison ends right there as

Buganda differs sharply from medieval England in many respects. While it was possible for persons of modest backgrounds to ascend to the highest pinnacles of society in Buganda, the same cannot be said of medieval England. On the question of protecting the rights of minorities and the underprivileged, Buganda was also ahead of medieval England. The monarchy in Buganda was sensitive to minority issues and concerns. Accordingly, it created avenues for addressing these concerns.

Hastings (1997: 156) draws attention to pre-colonial Buganda's ingenuity in national governance when he states that 'If there existed one nation-state in nineteenth-century black Africa, Buganda would have a good claim to be it.' He then went on to note that the centralized Kingdom of Buganda had been carefully organized into well-defined geographic areas or divisions for purposes of administrative efficiency and effectiveness.

The Kingdom of Buganda had made remarkable advances in other areas worth noting for the purpose of the present discussion. Long before the introduction of colonialism in the region, the kingdom had already instituted government revenue generating strategies, including a system of taxation that taxed all adults resident within the territory. The kingdom also had an army and a navy, in which all adult men in the territory were required to serve, and a well-integrated economy (Wrigley, 1996).

Tradition, administration and development

As the colonial era drew to a close in most of Africa in the late-1950s and early-1960s, members of the development community were compelled to wrestle with many crucial questions. Prominent in this connection was the question of how to cope with the wide range of complicated administrative quandaries which development planning inevitably elicits. Albert Waterston (1965), one of the gurus of development planning of that era, dedicated a whole chapter of his classic work, *Development Planning: Lessons of Experience*, to discussing 'administrative obstacles to planning.' Waterston noted that, in one emerging country after another, 'it has been discovered that a major limitation in implementing projects and programs, and in operating them upon completion, is not financial resources, but administrative capacity' (Waterston, 1965: 249). Today, three decades subsequent to Waterston's insightful observation, the situation remains unchanged as administrative ineptitude remains a leading, if not the most dominant hurdle to development in African countries. This assertion is hardly disputable. What fails to command consensus is how to go about eliminating this obstacle. What steps must be taken to improve public administration systems in these countries?

International and national development authorities seeking to address this question have historically ignored the need to evaluate African traditional models of governance with a view to identifying elements thereof that may improve the capacity of contemporary systems of administration on the continent. Yet, aspects of traditional and indigenous models of administration such as those relating to conflict resolution hold enormous promise for resolving some of the most nagging and complex disputes that either threaten to suffocate, or have culminated in irreversible damage to parts of the continent. I submit that the most notorious of the conflicts I have in mind, including but not limited to, the Tutsi/Hutu conflict in Rwanda, and the civil wars in Sierra Leone, Liberia and Cote d'Ivoire, would have been better dealt with within the context of a traditional African state. Also, although many in the West are either unaware or less likely to admit, the tyrants and despicably corrupt leaders that Africa has come be identified with in recent history, including Uganda's Idi Amin, Zaire's Mobutu Seseko, and Nigeria's Abacha, to mention just a few, would have been short-lived and would have caused far less damage to the economies of the countries they led in the context of traditional African statehood as opposed to a modern or more accurately, Western-type state.

One critical distinction between traditional African states and the so-called modern varieties that were shoddily constituted by European colonial powers in the late 19th century needs to be illuminated here. The colonial states, which form the foundation upon which the contemporary African state is reposed, were crafted to serve colonial development objectives, particularly exploitation of the colonial territories. In concert with this overarching objective of the colonial project, the administrative machinery of the colonial state was set up to essentially extract and repatriate raw materials from the colonies to the colonial master nations. Preparation for independence in the colonies did not include re-orienting the administrative systems of the emerging nations to focus on national development. Rather, the indigenous leaders that the colonial authorities hand-picked to inherit the colonial administrative machinery had to pass one litmus test, namely ensuring continuity.

Ensuring continuity meant that although in theory the new states were politically independent, they were to put the politico-economic interest of the 'erstwhile' colonial master nations ahead of the emerging nations' national interests. In exchange, the erstwhile colonial master nations guaranteed the hand-picked indigenous African leaders or what were essentially stooges of erstwhile European colonial powers in particular, and the West in general, political protection at all cost. This explains in large part the fact that efforts to replace internally unpopular regimes in natural resource-rich African countries have, more often than not, proved abortive. Most Africans agree that their leaders strive to serve not the interest of the citizens but their individual interest and that of Western powers. To the average African, the modern state is a tool designed to serve these two interrelated interests as opposed to the interest of the people.

The traditional African state differs markedly from the so-called modern variety. The King of contemporary Asante, the Asantehene Otumfuo Osei Tutu II, aptly described the fundamental basis and principles of the African traditional state in his keynote speech at the Fourth African Development Forum in Addis Ababa, Ethiopia on 12 October 2004 (Tutu II, 2004). As the king explained, traditional African societies have been organized throughout the centuries on the basis of a social contract in which people came together because they were convinced that such unification possessed the potential of guaranteeing them the attainment of a greater good. Such good was of the order and magnitude that members of the group as individuals or communities could not realize on their own.

At this juncture, I must pause to caution against the risk of indiscriminately exalting all threads of African traditional practices. To do so is as disingenuous as baselessly disparaging these practices. I have already characterized the basic structure of traditional African state. Under this structure, particularly in the case of centralized states, people agreed as a collectivity, to accord the king, queen or other similar leader, the power to make decisions for the greater good of society. Here, individual rights and privileges yielded to their communal equivalents. A few similarities between the traditional African state and the Western equivalent are discernible. Tutu, II (2004: 2) draws attention to one such similarity when he states that the traditional African state 'had all the elements of an Austinian state – a political sovereign, rule with his council of elders and advisors in accordance with the law.' He then went on to argue that,

> Although autocracy was not unknown, the rule of law was a cardinal feature of their system of governance. The king was ultimately accountable and liable to deposition upon the violation of norms considered subversive of the entire political system or particularly heinous. In many cases the political structures were complemented by hierarchy of courts presided over by the king, the head chief or the village chief (Tutu II, 2004: 2).

Thus, in traditional Africa, as in Western-style states, the citizenry agrees to have an entirely different entity, the state, act in their general interest. However, it is risky to make much of this apparent similarity. We cannot ignore the fact that in Western-style, particularly democratic states (as opposed to communist, authoritarian and cognate states), the chief executive is an elected official. In contrast, the chief executive – that is the post of king, chief or identical official – in the traditional African state is hereditary. Thus, in his speech to the Fourth African Forum (see above) the Asante King was overstretching the similarities between the traditional African governance structure and its democratic equivalent when he likened the institution of king-makers to the Electoral College in democratic polities such as the United States.

The skepticisms of critics particularly with respect to the ability of hereditary rulers whose legitimacy derived from the circumstances of their birth to act democratically and be accountable to the people are well-founded and in fact, healthy. However, the criticism that such leaders were despotic and operated under no legal or political control is somewhat unjustified. At any rate, it would be foolhardy to recommend monarchies as a solution to the inept and terribly corrupt institutions that have been wreaking havoc almost everywhere in Africa since the demise of colonialism. The fact, as Tutu II (2004) stated that, a king or chief is selected from an eligible pool of members within a ruling family does not make the option any more appealing or democratic.

Although there is no way of ascertaining this, it is quite possible that monarchical systems in Africa were eventually to evolve into democracies were it not for the destruction of indigenous institutions occasioned by the brutal assault of colonialism.

Lest we forget, most of the democratic polities of Europe today were, at some point in their evolution, monarchies. In some, such as the United Kingdom, the monarchies, which are mainly ceremonial today, function in tandem with the democratic institutions. Pre-colonial African societies, as mentioned earlier were predominantly non-centralized and contained elements typical of democratic institutions. Centralized polities were the exception and not the rule. Thus, an examination of governance structures in non-centralized can be very revealing and potentially beneficial for efforts designed to promote democracy throughout the continent.

In non-centralized or stateless traditional societies, such as the Metas of Cameroon, the Ibos of Nigeria, the Talensi of northern Ghana, the Sukuma of Tanzania, and the Nuer of southern Sudan, social control was ensured through the 'dynamics of clanship' (Tutu II, 2004). 'The normative schemes consisted of elaborate bodies of well-defined rules of conduct, usually enforced by heads of fragmented segments, and in more serious or subversive cases, by spontaneous community action' (Tutu II, 2004: 2).

Despite the absence of a highly centralized structure, traditional African stateless societies functioned under a well-defined set of sophisticated norms. Also, the decision making process in these societies benefited from the direct participation of the people as members of one of several societal groups – the extended family, clan, village or enclave. This, in essence, was democracy at its best. To the extent that this is true, it renders defenseless the popularly held belief that Africans have no tradition of democracy. Theories rooted in this belief have been advanced to rationalize the woefully disappointing record of ongoing efforts to institute democratic institutions in Africa.

Decentralization holds far more promise for development in Africa than its converse, centralization. The fact that the roots of decentralization run deep in African history cannot but be considered beneficial to efforts designed to improve administrative capacity throughout the continent. Before identifying and discussing the merits of decentralized administrative systems, I take a moment to examine the rationale for favouring highly centralized administrative structures on the part of colonial powers.

Colonial authorities had the primordial goal of resource-exploitation. Thus, whatever else was done, such as the creation of institutions for maintaining peace and order and those responsible for revenue generation, was only done as a means of facilitating attainment of this primordial goal. Centralized administrative systems assured tighter control over the colonial government budget, minimized cost, ensured consistency and accountability, and took advantage of the benefits associated with economies of scale and agglomeration. For colonial authorities, nothing besides making the most financial profit with the least overhead cost was important.

The literature on European colonialism in Africa is wont to err by associating administrative centralization exclusively with the French while presenting the British as favouring decentralization. A more critical read of the administrative strategies of the two colonial powers reveals that both were inclined to centralize authority. Thus, for instance, the British are on record for creating centralized chiefdoms where non-existed prior to the colonial era. The colonial Warrant chieftains that were created in British colonies exemplify this practice. Lest we miss the point, the kingdoms (e.g., in Yorubaland in Nigeria, Asante in Ghana, and Buganda in Uganda) and emirates (e.g., Hausaland, Nigeria) that were retained under 'indirect rule,' were already highly centralized polities before the colonial era. Thus, retaining them dovetailed neatly into the colonial philosophy of centralization.

Authorities in contemporary Africa must wrestle not only with questions of efficiency but also those having to do with effectiveness. To appreciate the distinction between these two important objectives of public administration, it is important to understand that efficiency is an internal measure, while effectiveness is an external performance measure. Efficiency deals essentially with questions relating to cost, while effectively concerns the extent to which the relevant public is satisfied with the public goods and/or services. To be effective, governments in Africa will need to be more knowledgeable and attentive to the needs of the citizenry. This invariably means decentralization, in other words, bringing the necessary public services as close to the citizenry as possible. Meaningful decentralization cannot be attained unless it is accompanied by devolution. Devolution entails empowering sub-national or provincial officials by redistributing authority, responsibility and resources for realizing public projects from the national to local level. As stated earlier, the roots of highly functional decentralized systems run deep in pre-colonial Africa and should therefore serve as fodder for ongoing institutional reform efforts on the continent.

Chapter 8

Traditional Resource Mobilization Strategies

Introduction

From pre-historic times, Africans have known nothing but a harsh and treacherous natural environment. Most of the northern portion of the continent is desert, while the central belt is characterized by humid tropical rainforests, lake basins, rapidly-flowing rivers that render river-based transportation difficult at best and impossible at worst, hilly, mountainous and rocky terrain that prevent the adoption of mechanized agricultural techniques, and dense forests that are replete with deadly insects and bacteria. In the hinterland areas – that is, the non-coastal regions – harvesting is done only once a year. Perhaps more worthy of note is the fact that Africans have always faced a scarcity of resources in locations where they are most needed.

The need to survive in this precarious environment partially explains the evolution of a variety of social, non-kinship common interest associations. I say partially because other traditional African non-kinship groups such as age-sets groups, which group individuals by age and gender, had as a primary objective promoting socialization. Of course, it is arguable that survival constituted one of the secondary objectives of these groups. Throughout Africa since pre-colonial times, common interest associations have played critical economic, social, political and cultural roles, which transcend and/or complement the usually more pervasive roles of lineage and clan. Four of these groups, rotating task execution group (ROTEG), rotating credit association (ROSCA), savings and loans club (SLC), and hometown association (HTA), which are mostly rooted in age-sets, kinship or lineage teams, command importance. The main purpose of this chapter is to promote understanding of these institutions, which have enormous promise for contemporary development efforts on the continent. It accomplishes this task by exploring each of the institutions in terms of its history, purpose, and nature. These institutions have proved not only resilient – surviving centuries of social and political changes – but also exceedingly dependable as vehicles for mobilizing and maximizing the utility of scarce and non-substitutable resources.

Rotating Task Execution Groups

A rotating task execution group (ROTEG) is an informal common interest group of two or more persons constituted for the purpose of rendering a specific service or set of defined services for each of its members in turn. These tasks are usually of the genre that each member working alone is incapable of completing in an effective and expeditious manner. Rotating task execution groups (ROTEGs) have evolved over centuries into what is popularly known as rotating credit associations (ROSCAs).

Historical background

Although as I argue below, rotating tasks execution groups (ROTEGs), the forerunner to rotating credit associations (ROSCAs) originated in Africa, I must concede that the prevalence of the latter throughout the developing and developed world alike renders my position somewhat shaky. Rotating task execution groups constitute a form of voluntary common interest association. According to Anderson (1971) horticultural villagers in different regions of the world, including Africa evolved a variety of such associations. The date of their emergence is unclear although Anderson (1971) speculates that it could have been between 8000 and 7000 BC. This period coincides with the establishment of the first agricultural communities in the Middle East. The uncertainty surrounding when and where the first voluntary association was created does not threaten the thrust of my argument, which is essentially that ROTEGs predated the arrival of Europeans in Africa by at a least a couple of generations. Therefore, I am on *terra firma* to contend that ROTEGs and ROSCAs are genuine African traditions. A practice does not have to be indigenous to a region to be considered a tradition of that region. After all, a basic litmus test for a practice to qualify as a tradition is that it would have had to endure through generations (Gyekye, 1997). Most practices considered traditions by many societies around the world trace their origins to somewhere outside of those societies. Thus, traditions are generally comprised of indigenous customs and elements of imposed or voluntarily adopted alien cultures.

As stated earlier, Africans have always lived in harsh environments. Thus, the need for cooperation has always been preeminent. Africans needed to cooperate as hunters and gatherers, and later as farmers. It was quite common for young men to embark on joint hunting expeditions and for young women to go fishing as a group. Over time, as Africans began to increasingly settle in enclaves such as villages, the need for building construction grew. Initially, the materials for the simple housing units they needed were readily available. However, as the design for these units grew more sophisticated, the concomitant need for complex materials obtainable from far away locations significantly increased. Accordingly, families working on their own could no longer build their own housing units. Consequently, groups originally created for the purpose of hunting and/or fishing assumed a new responsibility, namely constructing housing units for their members. Over time, the groups assumed additional responsibilities such as completing farming-related chores (e.g., clearing, weeding and harvesting), fetching firewood, and so on, for their members.

Nature

ROTEGs have always functioned as informal groups. Their informal nature notwithstanding, they operate based on a few, albeit unwritten rules. For instance, there is usually, at a minimum, some tacit agreement on what service has to be performed for which member and when. In most cases, the 'what' question is simplified by the fact that the same kind of service is performed for every member. These groups are usually of an ad hoc nature and are typically constituted during the clearing, weeding, planting and harvesting seasons. Thus, for instance, during the harvesting season, such a group may be established to harvest the crops of their members in rotation. However, all members do not need to necessarily request for the same service from the group. A member with a pressing need to mold building blocks may request the group to assist her/him in molding building blocks in lieu of harvesting crops. In any case, every effort is made to ensure equitability in the magnitude and quality of service rendered to each member of the group.

Membership

Two of the most important criteria that determine membership in ROTEGs, namely distance and age are worthy of note. A ROTEG typically draws its members from the same neighbourhood, village or town in which it is located. One important reason for this is that members need to be within close proximity to one another. In the pre-colonial époque the importance of proximity was amplified by the absence of effective and efficient transportation and communication facilities. A second important criterion is age. Each group is usually comprised of members of the same age group. This condition is important for many reasons, prominent amongst which are the following. First, it must be understood that apart from its economic function, ROTEGs play an important socio-cultural function. They reinforce social bonds amongst peers. Drawing members from the same age-group ensures that each ROTEG is comprised of individuals capable of performing at approximately the same energy level. The importance of this criteria cannot be overemphasized in view of the *raison d'être* of ROTEGs, namely to perform manual work for their members.

Operating procedures

A rotating task executing group (ROTEG), as stated above, exists to help its members complete tasks that cannot be effectively and/or efficiently completed by each member working solo. The standard operating procedure of a ROTEG is simple and can be summarized as follows. Consider a ROTEG comprising five members, A, B, C, D, and E, who agree to harvest the crops of each member for one day in turn. If the group begins by harvesting A's crops on Monday, they will be working for B on Tuesday and by Friday of the same week, *ceteris paribus*, they will be on E's farm. At the close of the working day on Friday, the group would have effectively completed one cycle. Thus, a cycle is deemed complete when each member has benefited once from the group's service.

Rotating Savings and Credit Associations (ROSCAs)

I have already drawn attention to the fact that ROSCAs evolved from ROSTEGs. Anthony Nwabughuogu (1984: 47) lends support to my assertion when in discussing the history of this institution among the Ngwa, a sub-group of the Ibo's in Nigeria, he stated as follows:

> The Ngwa have a tradition of pooling labour to break farm bottlenecks. Transition from labour pools to pooling money resources was easy enough. The Ngwa therefore evolved a credit institution – the *isusu* which became the main source of raising capital to meet their socio-economic needs in the pre-colonial period.

He then went on to advance a theory that situates the origins of *isusu* squarely in Ngwa, as opposed to having been borrowed from elsewhere. In this regard, Nwabughuogu (1984: 47) states thus, 'all sources indicate that *isusu* is indigenous to the Ngwa and had developed long before the establishment of colonial rule.' It is quite possible that ROTEGs evolved in different parts of Africa in response to local needs. However, whether it started in one part of the continent and was exported to other parts is not of critical importance here. What is of essence for the purpose of the present discussion is that the rotating credit association and other variants thereof are a genuine African tradition.

Despite dubious evidence to the contrary, ROSCAs existed in Africa in one form or another prior to the arrival of Europeans. Others (e.g., Rowlands, 1993) may beg to differ. Talking about the origins of ROSCAs in Cameroon, where they are commonly known as *njangi* or *tontines*, Michael Rowland opined that (Rowland, 1993: 80, citing Warnier, 1985),

> The *njangi* principle was developed in the palace societies, the members of which not only supplied goods and capital for trade but through the distribution of licenses such as the slave rope, controlled access to long distance trade in prestige goods.

It is true that some of the earliest and best-known efforts to systematically study ROSCAs and cognate institutions did not occur in Africa until the 1950s and 1960s (see e.g. Bascom, 1952; E. Ardener, 1953; S. Ardener, 1953; Gutkind and Southall, 1956; Isong, 1959; Geertz, 1962; S. Ardener, 1964). However, to consider these efforts as coinciding with the birth of ROSCAs is woefully erroneous. As far back as 1843, Crowther's dictionary of Yoruba vocabulary already contained an entry on rotating credit associations (RCAs), which it referred to in the local language as *esusu*. Shirley Ardener (1964) noted that a lot earlier than that, in 1794, to be exact, Sierra Leone already boasted a number of thrift clubs, operated in a manner akin to RCAs. This revelation casts a thick cloud of doubt over claims to the effect that Egba immigrants from Yorubaland in present-day Nigeria, introduced RCAs in Sierra Leone (see e.g., S. Ardener, 1964). Prior to the introduction of currency in Nigeria during the colonial era, the Yoruba's made *esusu* contributions in cowries. Also, there is evidence suggesting that ROSCAs existed among the Susu of Sierra Leone as far back as the 1880s when contributions were made in kind prior to the introduction of European currency (Ardener, 1964). In the ancient Kingdom of Bamum in neighbouring Cameroon, cowries were used for the same purpose before the introduction of modern currency by colonial authorities.

Researchers have also attempted to identify the factors that gave birth to RCAs or ROSCAs. In this regard, some have suggested that these institutions might have developed from a need to meet obligations such as assisting kinfolks in time of distress or with the payment of bridewealth (Nwabughuogu, 1984). Based on her study of RCAs in Cameroon in the early 1950s, Kaberry (1952) opined that they originated in palm wine drinking clubs. Others (e.g., Kuper and Kaplan, 1944) have suggested that these entities have their roots in the 'tea parties,' which were common amongst the Bantu-speaking groups of Southern Africa. Shirley Ardener (1964: 209), on her part, believes that RCAs evolved in response to 'the need to formalize uncodified traditional obligations as traditional sanctions were weakened by introducing the concept of regularity and rotation which distinguish these associations.'

Table 8.1   Sample of local names for informal savings and loans associations in Africa

CASE NO.	COUNTRY	COMMON LOCAL NAME(S)
1.	Benin	Ndjonu, Tontine
2.	Cameroon	Njangi, Djangi, Tontine
3.	Congo (DRC)	Ikelemba
4.	Egypt	Gameya
5.	Ethiopia	Ikub
6.	Gambia	Osusu
7.	Ghana	Susu, Akpee
8.	Liberia	Esusu
9.	Malawi	Chilemba
10.	Niger	Asusu
11.	Nigeria	Esusu, Isusu, Dashi, Adeshi
12.	Sierra Leone	Asusu
13.	Senegal	Tontine
14.	Sudan	Sanduk, Khatta
15.	Zimbabwe	Chilemba

Source: Based on data culled from Miracle et al. (1962).

The foregoing statements can best be classified as efforts to determine why RCAs emerged as opposed to where they originated. No researcher, to my knowledge, has traced the origin of a rotating credit association in Africa to a location outside of the continent. Rather, the RCAs functioning in locales as far away as Trinidad, and other Caribbean Islands, have been credited to slaves from Africa. Thus, I remain steadfast in submitting that the RCAs or other variants thereof are genuinely African traditions. As Table 8.1 shows, their prevalence throughout the continent is certainly not at issue. The table contains only a sample of ROSCAs and savings and loans clubs that have been studied since researchers began paying attention to such institutions in the 1950s. As I stated earlier, these institutions evolved from rotating task-executing groups (ROSTEGs). The transition from pooling labour and completing miscellaneous tasks for their members, to pooling other resources such as money was not only easy but also appears to have been a logical and rationale adaptation to a changing socio-economic environment.

However, definitions of the term ROSCAs in the relevant literature tend to be oblivious to their history and give the false impression that these entities exist to do no more than provide a source of credit and an avenue for saving to their members. For instance, Geertz (1962: 243), one of the first to illuminate the importance of rotating credit associations (RCA) or rotating saving and credit associations (ROSCAs), viewed the basic underlying principle of this entity to be as follows.

> A lump sum fund composed of fixed contributions from each member of the association is distributed, at fixed intervals and as a whole to each member of the association in turn.

The foregoing statement overemphasizes in-cash contributions while completely ignoring contributions of the in-kind genre. As noted above, ROSCAs originated as common interest groups designed to provide a variety of services to their members. It is true that ROSCAs have evolved since the pre-colonial époque. However, this evolution has entailed, in large part, broadening the scope of ROTEGs to encompass the provisioning of additional services such as savings and credit to their members. This explains the logic behind the name, rotating savings and credit association (ROSCA). Also, it is erroneous to state, as Geertz does, that, one member always receives the whole sum (Ardener, 1964). Shirley Ardener (1964) noted the deficiencies inherent in Geertz's characterization and proffered the following definition as one that is more reflective of the true nature and purpose of RCAs. A rotating credit association, Ardener (1964: 201) opined,

> [is] an association formed upon a core of participants who agree to make regular contributions to a fund which is given, in whole or in part to each contributor in rotation.

Ardener's definition, appears to be at the very general level, and therefore seeks to account for ROSCAs and all variants thereof throughout the world. In fact, Geertz's own research, which focused mostly on Asia (see Geertz, 1962), noted that contributions varied from one member to another. In Africa, however, it is save to contend that since the introduction of modern currency, ROSCA contributions have always been fixed and equally distributed amongst all members.

To the extent that this is true, van den Brink and Chavas (1997) erred in tailoring their definition of the phenomenon after the one proffered by Geertz (1962). Van den Brink and Chavas (1997: 746) define a ROSCA as,

> an association of men and women who meet at regular intervals, for instance, once a month, and distributes a lump sum of money to one of its members. The sum is made up of the variable or fixed contributions of each member of the association.

While this definition might have been suitable in Geertz's Asian context, it is out of place in Cameroon, which constituted the empirical referent of the study by van den Brink and Chavas (1997). These researchers paint a more accurate picture of the Cameroonian situation, which typifies what obtains in most parts of Africa when they described the functioning of the ROSCAs in their study in the following terms.

> 10 individuals could meet every month and each pays $10 into a pool. Thus, each month there is $100 in the pool. This is then handed over to one of the members. In the next they again pay $10 each and a different member of the group receives the money, and so on until every member has had his or her turn.

The definitional problem highlighted here stems from a lack of appreciation of the fact that ROSCAs usually operate in tandem with cognate common interest programmes. For instance, in Cameroon, membership in a ROSCA usually entitles one to membership in a savings club known locally as a *meeting*. A well-prepared article on 'informal savings in Africa,' by Marvin Miracle and his colleagues (Miracle et al., 1980) is one of the few works that have taken the time to distinguish savings and loans clubs (SLCs) from rotating savings and credit associations (ROSCAs).

Rules and procedures

Despite their informal nature, ROSCA members are bound by strict rules. In contemporary times, it is not abnormal to prosecute defaulters in formal courts of law. In the past, however membership was strictly controlled to ensure that potential defaulters are excluded. Additionally, the fact that members originated in the same neighbourhoods, villages or extended families meant that defaulters could be dealt with in-house. There was always the fear of being ostracized from a community, lineage or extended family. For a traditional African, few penalties were/are considered more severe than being shunned or banished by one's own community or kin. Even then, there were still cases of defaults. The defaulters belonged to two main groups (Nwabughuogu, 1984). The first group comprises those who defaulted before their turn for claiming the contributed funds came around. The second includes those who defaulted subsequent to claiming the funds. Those belonging to the first category were considered less of a problem than those falling under the second category.

In pre-colonial Ngwaland (Ibo, Nigeria) as Anthony Nwabughuogu (1984) noted, if a ROSCA member defaulted prior to receiving her own share due to unforeseen circumstances such as illness, that member was required to simply seek a substitute who would agree to take over her responsibilities in the association. In the event that defaulter was unable to find a willing and able substitute, the association would take upon itself to locate one. In any case, the defaulter must agree to reimburse the substitute, including all expenses the latter might have incurred, such as the cost of the palm wine and other articles that were used for the purpose of hosting the meeting.

As a rule, each member was/is required to host the meeting on the day he/she is receiving the funds. However, in the event that the defaulter simply decided to unceremoniously withdraw from the association for other than unforeseen reasons, the defaulter was severely dealt with. Fines of one chicken for each round defaulted were not unheard of. In one case in pre-colonial Sierra Leone, as Ardener (1964) reported, the defaulter was fined a whole cow. The threat of stiff penalties such as these, made the option of abandoning one's obligations as a member of a ROSCA extremely unpalatable.

Defaulters falling under the second category faced even stiffer penalties as they were considered a greater threat not only to the association but also to the community's stability. It was not unusual, as Nwabughuogu (1984) reported in the case of Ngwa, Ibo (Nigeria) for the offending party to be brought before the council of elders, village council or other higher traditional authority. Initially, the offending party was usually fined one goat for simply appearing before that body and ordered to go and fulfill his/her obligations in the association. If the offender refused to act as requested, he/she was then treated as a debtor, owing the amount he collected from the association minus his/her share of the contribution. Once the village elders or other cognate body reaches this decision, it invokes and applies the traditional law(s) dealing with debts against the defaulter. This sometimes entailed confiscating the offender's property to repay the debt or where the offender owns no property worth the value of the debt, he/she was sold into slavery and the proceeds used to repay the debt.

In other parts of Africa such as Cameroon, where van den Brink and Chavas (1997) examined a number of ROSCAs, the risk associated with loaning money is minimized by ensuring that less-known and newer members are placed at the bottom of the fund-receiving ladder. Thus, in the event that a borrower in this category defaults on the loan payment only very few members, those coming after him/her, are affected.

Savings and Loans Club (SLC)

A savings and loans club is a common interest group constituted to encourage saving amongst its members. Savings and loans clubs, also known as meetings, usually, although not always, operate along side ROSCAs. Members of meetings are encouraged to save as much as they can. Such encouragement is usually in the form of peer-pressure. No specific amount is stipulated. Thus, savings are usually a function of each member's means. However, every member is levied a fixed sum. A proportion of this sum, which is the same for each member, is deposited in a special account. It is from this account that funds are withdrawn to defray the cost of dealing with an emergency involving any member of the group. In Cameroon, the pool is referred to as a 'Trouble Bank.' Another proportion of the sum is set aside for entertainment, especially festivities at the end of the calendar year (usually at Christmas), when members are allowed to withdraw all or part of their savings. The cost of maintaining the association is defrayed from funds originating in interest charged on loans made by the association. Members have the privilege of borrowing money from the association. Non-members are eligible to borrow only and if only a member acts as a co-signer of the loan. In this case, the amount for which the borrower is eligible cannot be excess of the co-signer's total savings in the meeting. Meetings may raise funds to execute projects of interest to them by accepting work in return for pay from their members or other external entities.

Membership

Savings and loans clubs, like ROSCAs initially found their greatest utility as vehicles for facilitating the adaptation of rural immigrants from rural areas in urban centers. Here, I hasten to underscore the fact that contrary to popular opinion, urban centers pre-dated the arrival of Europeans in Africa. Typically, and as an extension of the kinship and lineage system for which Africa is well-known, individuals from the same rural neighbourhood or village living in the same urban center or metropolitan area, agreed to constitute an SLC. Thus, membership was restricted to individuals from the same rural village. This requirement served as a built-in device that deters dishonesty, corruption and all unacceptable social conduct. While this device was not successful in completely eliminating problems such as defaults on loans, it ensured their containment.

Hometown Associations

The extended family and ancestry rank very high on a traditional African's list of priorities. The concept of hometown as employed here is inextricably interwoven with the twin phenomena of family and ancestry. The concept of hometown constitutes a manifestation of the high regard and interest Africans have for the family as an important social institution. An individual's hometown – in the African context – refers to the place of origin of that individual's extended family. African tradition dictates that the umbilical cord of a child, no matter where the child was born, be buried in the child's ancestral home, in other words, the child's hometown. Similarly, the tradition requires that the remains of an individual be buried in that individual's hometown.

For Africans, the concept of hometown is synonymous with 'place of birth.' Thus, it is commonplace to find Africans whose identification documents contain as their place of birth locales other than those in which they were actually born. These locales are usually the birth places of their ancestors. Consequently, in patrilineal societies, all children in a family tend to identify with only one place, the one in which their father and/or grand father/great grand father were born, as their birthplace. This is certainly not the case in Western societies such as the United States, where a family may have three children with each of them identifying a different locale as his/her place of birth.

A hometown association (HTA) is a social organization comprising individuals sharing a common ancestry but living away from their ancestral land. As Honey and Okafor (1998) explained, hometown associations repose on two main pillars, ties of kinship and ancestry. They are invariably the product of migration and urbanization. The manure for the growth of this institution was provided by associational life, which constitutes a critical component of the African social structure. Hometown associations (HTAs) constitute the most visible manifestation of this structure.

Migration is one of the leading factors explaining trends in the growth and development of HTAs. Earlier on in the pre-colonial era, there were few human settlements that could be considered urbanized in terms of their sizes. Consequently, there were very few HTAs because the number of persons living away from their ancestral land was insignificant. With the introduction of colonialism and the concomitant adoption of capitalism as a favourite mode of production, many Africans had to leave their ancestral lands to work in the colonial mines and plantations. Examples of huge migrations trends occasioned by colonial capitalist economic development endeavours abound. For instance, in the southern African region, mine workers were generally recruited from neighbouring countries, such as Botswana, Zimbabwe, Malawi, and so on. In Ghana and Cote d'Ivoire, it was necessary to hire labourers in the coffee and cocoa plantations from across the international borders in Burkina Faso, Mali and Niger. In Cameroon, the migration occasioned by plantation agriculture was both internal and international as workers were recruited from the hinterland regions, such as the Northwest and Western Provinces and from Ibo and Ibiobio lands in Nigeria, to work in the colonial banana and oil palm plantations along the Atlantic Coast.

The further away from home people are, the broader their sense of 'hometown' becomes. Thus, individuals who are just a few kilometres away from home, may define their 'hometown' in more precise terms, such as a specific village. For individuals a couple of hundred kilometers away from home, 'hometown' may be defined in terms of an entire clan. Once in a foreign country, individuals from the same country tend to see themselves as one and usually register in the same 'hometown' associations. Thus, Malians working in cocoa plantations in Ghana may bond and constitute a common hometown organization even if they come from different villages in Mali.

To retain a sense of community, migrants from the same village, town or region in their area of origin began constituting clubs or associations. These clubs were initially set up to address the social and economic needs of the migrants in their host region. Thus, the clubs served an important welfare function and provided a safety net for their members. In keeping with African tradition, the clubs ensured that the remains of deceased members were repatriated for burial in their ancestral lands. Hometown associations registered success in several other areas, especially during the colonial era. Their accomplishments can be better appreciated when they are examined within a 'social network' theoretical framework. From this perspective, immigrants in any region are considered to be patently aware of the need to bond and strengthen links between their host community and their hometown or ancestral community.

This awareness provides the impetus for constituting HTAs. Once constituted, the HTAs quickly assume several responsibilities. Some of these are foreseen while others are unanticipated. Typically, HTAs provided a base that facilitated the transition into a new environment for new arrivals in the urban centers, mining or plantation communities. In this regard, the experienced migrants provided housing, information on securing employment, and survival tips to the new migrants. With the passage of time, the aims of HTAs were broadened to encompass economic responsibilities in the ancestral lands of their members. In recent times, HTAs have sought to promote not only social exchange, but they have also pursued low-scale development by using family remittances as a form of economic aid in their ancestral lands or hometowns.

Development implications

Until recently, international and national development authorities in Africa have ignored – and one may say, been oblivious to – traditional resource mobilization techniques in favour of Western or so-called modern varieties. Recently, thanks to research endeavours that are increasingly demonstrating their exceeding utility, authorities are beginning to acknowledge these techniques. However, unless the techniques are well understood, efforts to employ them are likely to fail. Thus, there is an urgent need to promote understanding of these techniques, particularly as a means of highlighting the importance of integrating salient and useful aspects of African culture and tradition in contemporary development endeavours on the continent. The remainder of this chapter seeks to contribute to efforts aimed at addressing this critical need. I hasten to note that rotating task execution groups (ROTEG), rotating savings and credit associations (ROSCAs), savings and loans clubs (SLCs) and hometown associations (HTAs) are different variants of voluntary common interest associations (VCIAs). The discussion that follows focalizes on VCIAs in general.

Voluntary common interest associations, particularly those discussed here are certainly not without blemish. Consider, for instance, the case of ROSCAs. Despite their incontestable strengths, they have been criticized for their inability to treat members equitably. To have a clearer picture of the biased nature of ROSCAs, consider a ROSCA with twelve members. Say, the members agree to launch the association in January with one member receiving the contributed funds at the end of the month, and the second one receiving at the end of February, and the third receiving at the end of March and so on, until the contribution is paid out to the last member (Number Twelve) in December. While this set up may appear equitable at first sight, upon closer examination, it turns out not to be. Note that the first recipient remained a debtor to the other group members for twelve months, while the last recipient served as a creditor to the rest of the group for the same duration – that is, twelve months. As for the other members, they moved from debtors to creditors in turn. To appreciate this problem from a slightly different perspective, it is necessary to recognize that the last recipient saved every month of the year but never received any interest on his/her savings. Another criticism of ROSCAs and other informal credit markets is that they are incapable of competing, and that whatever utility they may have is likely to diminish with the development of formal credit markets (van de Brink and Chavas, 1997).

As for hometown associations, they have also been the object of criticisms. For instance, they may foster inter-tribal cleavages and tribalism, both of which contribute to some of the most nagging barriers to economic, social and political development in Africa. When such associations have been incorporated in self-help development projects, as I documented elsewhere (see Njoh, 2002; 2003; forthcoming), they have been criticized for facilitating exploitation. In this case, it can be argued that projects executed by hometown associations, especially those executed with the support or acquiescence of the state, are designed to subsidize the worldwide process of capitalist accumulation (Burgess, 1978).

The criticisms of African traditional resource mobilization techniques tend to fade significantly once account is taken of their strengths. They hold enormous promise for the development aspirations of African countries. From the pre-colonial era when ROSCAs made it possible for their male members to secure the funds necessary to get married, ROSCAs have grown to become a critical tool of development in Africa. Rotating savings and credit associations can no longer be seen simply as what Clifford Geertz called a 'middle rung in development' in 1962. The role of these traditional institutions in the development process can be summarized under broad social, economic, political and cultural categories.

Social development implications of VCIAs

Voluntary common interest associations (VCIAs) have several merits with positive implications for social development in African countries. Below, I discuss five of these merits, including their ability to:

- promote communication and socialization,
- preserve and diffuse culture,
- foster control,
- address complex social needs of citizens,
- promote inter-group exchanges, and
- strengthen rural-urban linkages.

Voluntary common interest associations (VCIAs) are considerably versatile as instruments of communication and socialization amongst members of an extended family, village, clan or any other basic social unit. In the busy and complex life of urban centers throughout Africa, members of families, villages, and clans resident in locales away from their ancestral lands hardly come by time to get together. Consequently VCIAs create the *raison d'être* and forum for individuals to come together, interact and exchange ideas on a regular basis.

Voluntary common interest associations (VCIAs) also permit individuals from rural areas to replicate some rural conditions in distant urban or other milieus. In the cases where the migrant population is of an international genre, VCIAs tend to enable members of this population to replicate some of the conditions prevalent in their native countries in the host countries. The case of African VCIAs in Europe, North America and other Western regions is illustrative. Within Africa itself, instances of international migrant workers, such as Malians in cocoa plantations in Ghana and Cote d'Ivoire; or migrant mine workers from Botswana, Namibia, Zimbabwe in South Africa, also exemplify this situation. It is difficult to overstate the importance of VCIAs for the social and psychological health of migrants in any community.

The importance of VCIAs is amplified in African countries with huge territories such as Nigeria, Sudan, and the Democratic Republic of Congo. The poor state of transportation facilities in Africa accentuates the need for VCIAs even in the smaller countries. Given the conditions that prevail in Africa, VCIAs tend to provide a 'home' away from 'home' for their members. In a study of Ibo's in Lagos Nigeria about half a decade following the country's independence, Kenneth Little (1965) arrived at a similar conclusion. At the time, Little observed that Ibo's in Lagos belonged to 'meetings' that roughly corresponded with the basic social units – village, village group, and clan – in their ancestral lands. By so doing, the Ibo's effectively transplanted their culture to a foreign land, namely Lagos. In his study of migrants in Bamako, Mali, Claude Meillasoux (1968) also observed that VCIAs served not only as tools of socialization but also as vehicles for the preservation and diffusion of culture. In other words, VCIAs served as tools for safeguarding and extending village life into the urban milieu.

Africa is certainly not unique in this regard. In fact, as Anderson (1971: 216) observed, 'peasant Ukranians who had migrated to towns and cities in France, for example, transplanted village customs as the activities of multipurpose associations in the host country.' In the United States, Hispanic voluntary associations have been exceedingly successful in ensuring the preservation of Latino culture while diffusing or promoting same in their host country. Voluntary common interest groups such as hometown associations (HTAs) and rotating task execution groups (ROTEGs) as well as rotating savings and credit associations (ROSCAs), also have a social control function. Voluntary associations, their informal status notwithstanding, have stringent rules that govern the behaviour and conduct of members. These rules are designed to reward approved behaviour and punish unapproved conduct. By so doing, these associations contribute in no small way to deterring immoral, delinquent and/or criminal behaviour.

Voluntary common interest associations have been successful in meeting the needs of citizens that the state in peripheral society is either incapable or unwilling to address. For instance, they help ease or facilitate the integration of rural migrants in urban centers throughout African countries. Usually, these groups strive to socialize new immigrants to new techniques and skills necessary for success in urban centers or their host locale in general. Voluntary associations are also potent as instruments for facilitating inter-group, inter-family and inter-tribal communication and unity in, especially urban settings. The demands of urban life in Africa are so stringent that individuals, even members of extended families, are unlikely to make time to get together and socialize. However, membership in voluntary common interest associations such as ROSCAs, SLCs, and ROTEGs compels them to meet on a regular basis. By so doing, VCIAs help to foster interaction and bonding among members of extended families, clans or villages living in communities outside of their ancestral lands. Furthermore, VCIAs foster inter-group linkages, interaction and understanding. To appreciate this capability of VCIAs, it is necessary to understand that members of VCIAs simultaneously belong to a multitude of other organizations, such as churches or other religious bodies, places of employment, educational institutions, and so on. Thus, members of VCIAs serve, by default or design, as ambassadors of their associations to the other organizations in which they belong.

Economic implications of VCIAs

The 1960s marked the end of the colonial era for most African countries. This same decade and the one before are best remembered for what did not happen as opposed to what happened. According to one of the leading authorities on the economics of development in peripheral countries, Gunnar Myrdal (e.g., 1956), this is what was supposed to happen. A variety of institutions designed to not only encourage savings, but also serve other important needs of their clientele in particular, and the surrounding communities in general, were supposed to have been established. The institutions, according to Myrdal's vision, were supposed to have been tailored in such a manner as to make them adaptable to different individual needs and possibilities, fit into the community patterns, and seek to promote planned and 'goal-directed' savings. Today, more than four decades subsequent to independence in most of Africa, Myrdal's laudable goals are yet to be realized. Why? This question has pre-occupied policymakers, economists and development planners of all stripes since the 1950s and especially since the 1960s. For modernization theorists, the answer is simple. Recognizing the need to save requires a fundamental change in attitudes on the part of the people of underdeveloped nations. Africans have not made this fundamental change and have therefore never recognized the need to save.

As I have demonstrated in this chapter, modernization theorists err, particularly by insinuating that African culture and tradition are incompatible with saving. It is true that economic development depends to a large degree on aggregate national saving, which must be increased to cover investment growth. While development economists of the 1950s and 1960s appeared aware of this, they appeared oblivious to, or dismissed off-hand, time-tested traditional methods of saving and mobilizing resources. Instead, economic development planners of the time concentrated exclusively on savings strategies that have proved successful in Western economies. These strategies include taxation, compulsory sale of government securities, economies in government expenditures, and cognate fiscal measures.

These constitute the hallmark of the Britton Woods institutions' structural adjustment and consanguine programmes that were initiated in these countries in the 1980s and continue to be encouraged in varying degrees throughout the continent. These strategies are inherently flawed, especially because they fail to achieve the goal of creating a propensity on the part of individuals to save – a goal that the traditional methods such as ROSCAs and SLCs have a time-tested record for achieving. Formal efforts on the part of policymakers and economic development planners to encourage individual savings in African countries have thus far gone awry largely because they have focalized on westernizing Africans. In this regard, they have attempted to change the general pattern of mores and social structure through the following strategies (Geertz, 1962: 242):

- propaganda campaigns aimed at encouraging individuals to purchase government securities; and
- creating Western-type savings institutions such as banks, savings cooperatives, credit unions and so on.

The observation by Clifford Geertz in the 1960s that these efforts have failed to yield any positive results remains true to-date. Geertz (1962: 242) proffers two explanations for this abysmal failure. The first is that,

> deep-rooted customs [of the native] yield very little to official sponsored exhortations to discard them in the interest of progress; and the second because the impersonality, complexity, and foreigness of the mode of operation of such 'capitalist' institutions tends to make traditionalistic peasants, small traders, and civil servants suspicious of them.

Rotating savings and credit associations (ROSCAs), savings and loans clubs (SLCs), and rotating task execution groups (ROTEGs), which all take on an identity akin to hometown associations (HTAs), play other familiar roles in national economic development. Individuals away from their ancestral lands are usually in need of funds for emergency purposes or as start-up capital for new business. Formal banks are, more often than not, inaccessible to these individuals. In such circumstances, migrants are likely to do well by simply seeking assistance from a ROSCA comprising members from his/her hometown (see for example, DeLancey, 1987, 1977). The conditions for borrowing are typically not very rigid. All an individual may need is either to be a member or have a co-signer who is a member. Thus, it is safe to conclude that as sources of loans and credits, ROSCAs and cognate traditional institutions are far more accessible, more personable, and more flexible than formal financial institutions. Also, in contrast to banks, members of ROSCAs and similar entities are not required to pay interest on loans. In other words, members have access to interest-free loans. Also, it is important to note that traditional financial institutions are far more effective and efficient than their counterparts of the formal sector. In this regard, it is necessary to note that traditional institutions usually do not have a separate staff of officers. This dramatically reduces or eliminates any cost associated with maintaining a paid staff of officers.

A further merit of traditional financial institutions can be appreciated by recognizing the fact that these institutions typically deposit their funds in formal banks. Thus, despite their informal status, traditional financial institutions are connected to the formal sector of the economy. These institutions are connected to the formal sector in several other ways. For instance, when individuals borrow from traditional financial institutions for the purpose of funding a business venture, the individual winds up paying a business tax and making other necessary payments to the state.

Yet another way of appreciating the economic importance of traditional institutions is by examining the contributions that foreign-based HTAs make to the economies of their home countries in the form of remittances. It is not important for the purpose of this discussion to separate the remittances made through HTAs and those made through alternative channels. Suffice it to say that remittances especially from the developed to the developing world have been growing tremendously in recent years. This growth has accompanied the sharp rise in the number of persons, especially professionals, from the south to the north. One positive upshot of this development is the fact that remittances from abroad have risen to levels comparable with development assistance and foreign direct investments in some African countries. Apart from remittances, individuals resident in foreign countries have always worked through their HTAs to significantly contribute to development efforts in their homelands. In particular, these HTAs have conducted book drives, collected hospital equipment and other necessary supplies, which they have shipped to their homelands. Additionally, HTAs have single-handedly completed or contributed to the completion of public infrastructure development projects in their homelands.

African governments will therefore do well to create a conducive environment within which HTAs can operate. More especially, these governments must treat HTAs and cognate indigenous entities as the partners in development that they are.

Chapter 9

Traditional Healthcare and Healing Strategies

Introduction

The importance of health as a factor in African development is incontestable. What is controversial is how to go about ensuring good health for the people of the continent, especially those in the sub-Saharan region. This region is turgid with some of the world's most nagging health problems. The deadly nature of these diseases is hardly news to those in the international development arena. What may be newsworthy is the fact that residents of contemporary Africa die or are permanently incapacitated by preventable or easily curable diseases such as malaria, diarrhea, dysentery and sleeping sickness.

It is true that modern medicine has gone a good way in eradicating some of the continent's health problems. In this regard, efforts to combat a number of deadly diseases such as malaria, syphilis and gonorrhea have registered considerable success. However, the cost of modern medical care for these and other diseases is prohibitively high throughout the continent. Thus, national and international health authorities will do well to entertain alternative healthcare strategies. One alternative that appears particularly promising is traditional medicine. By traditional medicine, I am specifically alluding to the practices and knowledge regarding healthcare that pre-dated the introduction of Western or so-called modern medicine in Africa. Although often appearing as no more than footnotes in Eurocentric historical accounts of Africa, Africans had developed sophisticated healing methods and techniques centuries before the arrival of Europeans on the continent. These methods and techniques were unique in many ways, but particularly because they were defined by African tradition, custom and beliefs. In addition, they were contextually relevant and adapted to the local environment.

Traditional medicine is only beginning to receive serious attention from modern (read, Western) medical experts. This is despite the fact that the healthcare needs of most people in Africa are met by the informal medical healthcare sector. The World Health organization (WHO) estimates that non-conventional healthcare systems address the healthcare needs of 80 per cent of all Africans (WHO, 2002: 1). Even more important is the fact that traditional healers outnumber conventional doctors throughout Africa.

If nothing else, this underscores the urgent need to incorporate traditional healers and medicine in ongoing efforts to address healthcare needs on the continent. Some have gone so far as to recommend that conventional medical practitioners co-operate with traditional healers as a means of enhancing the effectiveness of healthcare delivery systems in African countries (see e.g., Courtright et al., 2000). Courtright and his associates, rationalize this recommendation in a study focusing on collaboration between modern and African traditional healers in efforts to prevent blindness in Malawi, Zimbabwe and Nepal. They argued that (Courtright et al., 2000: 3),

- Traditional healers can serve as a bridge between community and district eye care providers;
- Some traditional healer practices, such as face washing, are useful in efforts to cure blindness;
- Traditional healers will always constitute part of the health policy landscape of Africa and other developing regions.

This tacit endorsement of traditional healing practices is of recent vintage and likely to be rejected by most in the formal medical field in particular and the development community in general. The tendency to reject traditional African healthcare strategies is in line with the well-known proclivity towards devaluing African achievements. Yet, available evidence suggests that Africans accomplished a lot in the medical field before the European conquest. Failure to incorporate indigenous accomplishments in contemporary efforts to deal with Africa's health problems explains, at least in part, the abysmal results that have been registered in this regard. The task of incorporating indigenous health strategies in contemporary healthcare practice is rendered difficult by the absence of sufficient knowledge on these strategies. There is therefore a critical need to expand the extant scope of traditional healing strategies. My aim in the present chapter is to contribute to efforts designed to address this need. I am not a medical expert and therefore leave it to others better qualified than myself to judge the scientific validity of any African traditional health knowledge and claims presented here. My aim is to consider some of the social implications of the strategies, especially those that were deemed to be of no value, and in fact considered to be antithetical to development efforts on the continent by early European visitors, colonial authorities and missionaries.

Traditional healing practice and its critics

Africa has the longest history of healing in the world and African traditional healthcare techniques and methods have spread through time, space and cultures. Maintaining their unique character, these techniques and methods have provided effective healthcare to a majority of Africans for centuries. Today, despite aggressive propaganda campaigns designed to promote Western or so-called modern medicine, traditional African techniques continue to be the healthcare option of choice for most Africans on the continent and in the diaspora.

Despite their popularity, traditional African healing practices and medicine continue to be the subject of virulent attacks and criticism. Some of the criticisms focalize on the fact that African traditional medical practitioners are at once 'priests, magicians and interpreters of dreams.'

Furthermore, skeptics deride African traditional physicians for not employing 'scientific methods.' What is more noteworthy is the fact that critics of traditional African methods and techniques have often been reluctant to prove that they are less effective than Western varieties. Rather, whenever critics are presented with incontrovertible evidence attesting to the effectiveness of traditional approaches, they resort to making every effort to show that the traditional technique in question has foreign roots. For instance, one critic of African traditional medicine tried to credit Greek health practitioners with the fact that Egyptian health resorts were enormously popular during the Graeco-Roman period (Todd, 1921). The only evidence offered by this particular critic is the fact that Greeks and Romans had been frequenting Egypt at about the same time (Todd, 1921: 461). It is puzzling that from records clearly portraying Greeks and Romans as visitors to Egypt subsequent to Alexander's conquest, someone would infer Graeco-Roman influence on Egyptian medical practices.

Yet, according to Herodotus who had traveled to Egypt about the middle of the fifth century B.C. when the country was still under Persian rule, Egyptian medical practice had advanced to the point where specialization was already in vogue. As Todd (1921: 462) himself quotes from an 1891 publication, 'Each physician applies himself to one disease only, and no more.' Von Klein (1905) also observed the extent to which medicine was specialized in ancient Egypt and contended that such a development helped to advance medical knowledge. Apart from the fact that specialized medicine was already practiced in Egypt long before this became fashionable in Europe and other parts of the world, there was also an abundance of physicians in the country. There were physicians for the eyes, head, teeth, belly and internal disorders. Critics are however, not convinced by the abundant evidence brought forward to bolster these claims. Todd (1921), for instance uses the fact that one of the earliest records attesting to the presence of an eye doctor was written in Greek to argue that Greeks would have had a hand in the development of such a specialization. Thus, without any attempt to substantiate his claim, Todd (1921: 462) contended as follows.

> It must be emphasized that during the later period of Egyptian history, the Egyptians were probably much influenced by the Greeks and Romans and that true specialization in medicine was of Greek origin.

Notice the quantum leap from speculating that, 'Egyptians were probably much influenced by the Greeks and Romans' to the conclusion that 'specialization in medicine was of Greek origin.'

Africans recognized the cleansing powers of water and were well-known for their use of water for therapeutic reason, that is, hydrotherapy. Thus, enemata and douches were commonly used as a remedy and/or cure for a number of ailments. However, critics charge that, 'hydrotherapy was a fetish' for Africans (see e.g., Todd, 1921: 462). Todd (1921: 462) went on to argue that, 'a physician whose function it was to cure by enemata was an exponent of a cult rather than a specialist as we understand the term today.' Paradoxically, he acknowledges the fact that some of the herbal cocktails that were developed in Africa, particularly in Egypt were later adopted by Greeks and were still present in modern pharmacopoeias in the 1920s when he was writing. This is despite the fact that he branded the cocktails as 'collections of incantations and weird random mixtures of refuse with roots and other substances' (p. 463).

What critics branded 'hydrotherapy,' and derided as a 'fetish,' is, first and foremost part of the preventative measures against disease that Africans have employed for centuries. Maier (1979) quotes a number of early European visitors to Africa who had been amazed by the religious attachment of Africans to the basic principles of hygiene. For instance, in Asanté, which was a well organized and highly centralized polity, strict municipal rules required all streets to be regularly cleaned. Also, individuals were required to burn all rubbish and offal from their homes every morning. Such burning had to be carried out in designated spots at the back of the streets. The Public Works Department of the ancient kingdom was in charge of enforcing sanitation and hygiene laws. In this regard, relevant officials from the department cleaned the streets and ordered citizens to do same for their immediate surrounding on a daily basis.

One cannot but notice the remarkable similarity between the use of sanitation as a disease-prevention strategy in pre-colonial Africa and the sanitary reform measures that were later implemented in Europe in general and England in particular in the mid-1800s. Here, it is worth recalling that before the revelation that a significant number of diseases have their roots in viruses and bacteria, diseases were believed to be the product of noxious air. The theory of miasma – a Greek word for pollution – was popular within the European medical science community as recently as the 19th century. This theory, whose roots are traceable to the Middle Ages, was used to explain the cholera epidemic of London in the 1800s. Thus, up until the mid-1800s Medical and cognate scientists believed that dysentery, diarrhea, cholera and later, malaria were air-borne diseases. From this vantage point, purifying the air and eliminating all sources of foul or pungent odours could prevent disease.

Although the miasma theory was rendered absurd by later scientific work, which more convincingly linked diseases to bacteria and viruses, it constituted the basis for the important sanitation reforms that took place in Europe in the 1800s. It also provided the solid foundation upon which modern principles of hygiene were built. In this regard, the work of Florence Nightingale (1820–1910), the Crimean War nurse, who pioneered efforts to keep the air in hospitals fresh and clean, is especially remarkable.

Therefore, the efforts of pre-colonial Africans who saw a direct connection between sanitation and disease must be judged in light of the knowledge of medical science that was available at the time. Sanitation conditions continue to be among the leading determinants of health status, notwithstanding microbiological findings to the effect that microorganisms, which grow by reproduction, create diseases.

Another potent and certainly sophisticated disease prevention strategy known as variolation, entailed blowing dried smallpox scabs into the nose of a healthy person. This person then contracted a mild form of the disease and upon recovery, was immune to smallpox (Maier, 1979). This technique had actually been developed in Asia but was adopted by healers in Africa as far back as 1700 – more than a century before the onset of the colonial era. The technique was already commonplace and exceedingly effective in Asante by 1784 (Maier, 1979). Long before variolation, ancient Egyptians had made considerable advances in the area of surgery. However, critics (e.g., Todd, 1921) have been quick to discredit any advances in surgery by traditional Africans. Africans, critics have contended, were capable of doing no more than incising abscesses and performing ritualistic circumcision. Yet, there is evidence suggesting that pre-colonial Africans carried out orthopedic surgery. For instance, as revealed by studies of mummies of the period, rude splints were employed in Egypt in cases of compound fracture of the femur and forearm as far back as the Vth Dynasty (circa, 2600 B.C.) (Smith, 1908). A lot later in the 19th century, a number of traditional African doctors were reputed for practicing corrective medicine in Asante, Ghana (Maier, 1979). The procedure used by these doctors was simple and comprised binding a fractured arm or leg in a splint (just as the ancient Egyptians did).

Dental surgery is another area in which pre-colonial Africans in general and ancient Egyptians in particular were relatively skillful (see e.g., Frazer and Downie, 1938). Studies of Egyptian mummies have revealed the presence of mouthpieces (e.g., Todd, 1921). Yet, critics have been relentless in discounting the achievements of Africans in this area. According to some of these critics (e.g., Jones, 1907), there are no data suggesting that the teeth of anyone have ever benefited from a dentist's handiwork in ancient Africa. Others go as far as to make the bold and erroneous claim that ancient Africans could 'not even extract teeth, let alone perform operations calling for considerable knowledge and skill which such a procedure would imply' (Todd, 1921: 468).

Dental surgery including the extraction of teeth, and the extraction of all or parts of a single tooth for health or cosmetic reasons had been undertaken in Africa long before the arrival of Europeans on the continent. The discovery of a number of teeth bound together with gold wire in a tomb in Egypt (Ruffer, 1921) suggests that dental braces for cosmetic or other reasons were known in ancient Egypt. Early European researchers, who observed the practice of teeth extraction and/or shaping in sub-Saharan Africa, labeled it 'tooth mutilation' (see e.g., Goose, 1963; Frazer and Downie, 1938). It would appear that the researchers who have been attracted to studying this practice were more intrigued by the seemingly exotic purposes rather than the medical or therapeutic rationales of the practice. In discussing this practice, N.W. Thomas (1916) stated that the 'mutilation' of teeth by shaping or laxation in Africa is carried out for a number of reasons, including making the men 'look more warlike' or 'for decoration to attract the opposite sex.' According to Frazer and Downie (1938), the Damara of South Africa knocked out their front teeth to enhance their ability to speak their native language. Among some of the Ibo of Nigeria as Thomas (1913) noted, this practice constituted part of the initiation process. Here, women were allowed to bear children only after their teeth had been filed. The only medical reason for tooth extraction or shaping that was reported related to the Massai of East Africa. For the Massai, according to D.H. Goose (1963), extracting a tooth or part thereof to create a gap in the front portion of the mouth was necessary to facilitate food intake in the event that a person were victimized by a disease such as tetanus that prevents the mouth from opening.

The fact that none of the afore-cited studies treated so-called 'teeth mutilation' as dental surgery, does not mean that dental surgery was unknown among pre-colonial or ancient Africans. Rather, this affirms the tendency of a significant number of European and other Western researchers to concentrate on the seemingly sensational and often mundane, rather than the serious and more laudable aspects of the African experience.

Traditional Africans were also adept in other aspects of healthcare. In describing the traditional African process for treating dysentery in the early-18th century, Henry Meredith stated that the therapy consisted of purging the patient's bowels with purgatives, introducing astringent and stimulating clysters into her system, keeping her warm, frequently embrocating her loins and belly with a composition of pepper, and using suppositories (Meredith, 1812: 243). Evidence suggests that Africans also possessed sophisticated skills and talents for dealing with several other ailments. In this regard, Africans had developed herbal and vegetable remedies for use in combating some of the endemic diseases of the region long before the European conquest. According to an early European visitor to Asante, medications from natural products such as lemon and lime juice, malaget (also known as Grains of Paradise or Cardamon), the roots, branches and bums of trees, and a litany of green herbs, which were impregnated with an extraordinary sanative virtue were widely used in the 18th century (Maier, 1979; Bosman, 1705). Certainly more deserving of attention here is the observation that,

> the green Herbs, the principal remedy in use amongst the Negroes, are of such wonderful efficacy, that 'tis much to be deplored that no European Physician has yet applied himself to the discovery of their Nature and Virtue (as quoted in Maier, 1979: 66).

To place this in some meaningful context, it is necessary to recall that European medicine of the time comprised largely of 'bleeding, cupping and using laudanum, purgatives, emetics and salves' (Maier, 1979: 66).

Pre-colonial Africans had also developed skills, techniques and methods for treating some of the diseases that continue to menace Africans to date. Fevers, for instance, were commonly treated by frequent ablutions with warm water, and then rubbing over the body with certain herbs. In cases where the fever was accompanied by headaches and/or aching joints, a cocktail of pepper, lime juice and other herbs, was applied. Completing the treatment entailed using the bark of a tree, which bears a striking resemblance to the cinchona tree from whose bark quinine is made

(Meredith, 1812). The bark of the tree in Henry Meredith's meticulous account was harvested from the northern part of the Gold Coast. A re-examination of Meredith's account strongly suggests that the tree in question is what came to be widely known as 'African quinine' *'(Crossopteryx febriguga)*, which contains a glucoside B-Quinovine (found in other Cinchoneae)' (Maier, 1979: 67). Treatment for some eye-related ailments consisted of 'drawing blood from the temples and forehead and by dropping limejuice into the eye' (Maier, 1979: 67).

As the above quotation suggests, a good many Europeans who were taken ill while in Africa had occasion to benefit from African healing knowledge. In a number of instances, as can be gleaned from the following true story, some indigenous African herbs and healing techniques were transplanted to Europe (see HerbalAfrica. com). The story has to do with an English tuberculosis patient by the name Charles Stevens, who had traveled to South Africa in 1897 in search of treatment. Once in South Africa, he was directed to an African traditional doctor of Basuto extraction. The doctor prepared and gave a decoction of a local medicinal plant. The decoction was specifically of the roots of *Pelargonium sidoides*, a species of geranium, which is indigenous to South Africa. The Zulu refer to the decoction as *'umKhulkane'* (denoting respiratory infection) and *'hHlabo'* (which roughly means chest pain). The results of the Basuto doctor's therapy were exceedingly positive, as it did not take long for Mr. Stevens, the tuberculosis patient, to be cured. Once fully recovered, Mr. Stevens transplanted to England the *pelargonium sidoides*, the 'mysterious plant' that had been used by the traditional African doctor. Once the plant had grown and matured, Stevens was able to duplicate the decoction, which had been prepared for him by the traditional African doctor in 1897. Steven's decoction became very popular in Europe under the name, 'Steven's Consumption Cure.' A former missionary medical doctor named Adrien Sechehaye, discovered Steven's 'Consumption Cure' in 1920 and made a record by curing over 800 tuberculosis patients in Switzerland with a homeopathic preparation of this decoction from 1920 to 1929. The advent of synthetic tuberculosis drugs contributed to diminishing the popularity of Steven's remedy in the medical field in Europe until it was 'rediscovered' a few years ago.

Contemporary validation of traditional herbal healing

Those who advocate 'modernization' for its own sake are often oblivious to the fact that tradition is certainly not synonymous with ineffectiveness and inefficiency. Thus, the important question that must be addressed before discarding any traditional practice in favour of a 'modern' substitute is as follows. Which of the two methods is more effective (i.e., able to address a defined need) and more efficient (i.e., able to address a defined need at an affordable cost)? To the extent that the necessary herbs and herbal healers or herbalists are locally available, it is safe to assume that any given traditional herbal healing option is likely to be less expensive than its imported competitor. Thus, the question with which I concern myself here is the former. Are the traditional herbal healing strategies effective? A cogent response to this question has been proffered by Anne McIntyre, a practicing medical herbalist, and past director of the National Institute of Medical Herbalists. In a well-written and persuasively argumentative paper for Positive Health's Complementary Health Magazine, *Positive Health, Online*, McIntyre pointed to the growing and proliferating literature on herbal healing as a sign that traditional medicine is making a come-back. She went on to state that,

> Recent research not only validates their [herbs'] ancient medicinal uses but also takes this a few steps further by helping us to understand the biochemical mechanisms involved.

To address this question, I turned to two very informative internet sources, *'Positive Health,'* and *'Herbal Africa'* which are dedicated to health matters (see positivehealth.com; HerbalAfrican.com). Of particular interest for the purpose of the present discussion are the statements contained in the contributions from highly qualified medical doctors and other medical scientists regarding herbs that have been used by traditional healers in Africa for centuries. My description of the following herbs and their functions as employed by Africans and people from other regions are inspired by the said contributions.

- Sutherlandia;
- Eleutherococcus Senticosus Maxim Shrub;
- Devil's Claw;
- Kigelia Africana;
- Tisane;
- Neem (Azadirachta indica A. Juss);
- Purple coneflower (Echinacea purpurea/angustifolia);
- African Cucumis;
- African Ginger;
- Artemisia;
- Gladiolus;
- Helichrysum; and
- Hibiscus.

Sutherlandia

Traditional healers in Africa have used this herb to treat immune dysfunction, viral and bacterial infections, stress and depression, chronic fatigue syndrome, diabetes, rheumatoid arthritis, and menopausal symptoms. According to Dalvi (Online), Sutherlandia, which goes by a number of African names, each of which characterizes the plant as a curing agent and potent purifier of human blood as well as an all-purpose tonic, contains the following active compounds:

> L-canavanine, a potent non-protein amino acid and L-argenine antagonist with documented anti-viral, anti-bacterial, anti-fungal and anti-cancer activities. The second is pinito, a naturally occurring sugar which may stimulate glucose transport into isolated muscle cells

and which seems to have a clinical application in treating the neurotransmitter which is a safe nutritional relaxant and anxiolitic that can be used to control seizures and depressive feelings

Eleutherococcus Senticosus Maxim Shrub

This is an adaptogen that has been successfully used to enhance the body's ability to cope with internal and external stressors. The herbalists' claim in this respect was bolstered by studies that were conducted in Russia some thirty years ago. The studies revealed that the roots of this shrub contained ingredients that are capable of boosting the body's ability to fend off respiratory tract infections, viruses and influenza-type illnesses. Laboratory tests have shown that the said ingredients are capable of increasing the number of immune fighter cells, particularly T-Helper cells. Other studies have suggested that the ingredients in *Eleutherococcus* are capable of combating the incidence and severity of Herpes viruses, which are responsible for cold sores and genital Herpes infections (Williams, Online). Furthermore, the ingredients have been found to facilitate the process by which oxygen is transported to the muscles. Therefore, this particular herb is potentially useful for those functioning in extreme climates, cosmonauts whose natural systems may need adjusting to weightlessness and re-adaptation to gravity.

Devil's Claw

Another plant that has been used for therapeutic purposes by traditional healers in Africa is known as the 'Devil's Claw.' This perennial plant whose therapeutic value has been confirmed by a number of Western trained medical scientists (see e.g., Stannard, Online; Leith, Online) is commonplace in the southern and eastern parts of Africa, particularly Namibia and the former Transvaal region. Devil's claw, translated from the German word, *Teufelskralle*, has been in use for centuries in the southern region of Africa. It is from here that a German farmer by the name, G.H. Mehnert, transplanted the plant to Europe subsequent to observing its effective use by traditional doctors in Africa. In Africa, it has been, and continues to be used to treat arthritis, rheumatism, fevers and headaches. Additionally, it is processed into an ointment, which is then used to treat sores, boils and ulcers. The ointment is also used to relieve pain during labour and child delivery. Presently, the plant is an exceedingly popular commodity among smugglers who clandestinely transport it to Europe and other parts of the world. In Europe it is used to treat pain or other problems associated with the musculo-skeletal and digestive systems (Stannard, Online). Also, as a validation of African herbal knowledge, ingredients obtained from the plant are commonly found in medications for different genre of arthritis, rheumatism, and lower back pain. A number of studies have shown that Devil's Claw has enormous potential as a substitute for pain medications such as Ibuprofen and Brufen that are currently in use in Europe and North America. As reported by Liz Wickham (Online), Health Correspondent for *Carlton* and *London Tonight*, results from placebo-controlled trials in Germany have shown Devil's Claw to be a very useful substitute for conventional pain medications.

Kigelia Africana

The *Kigelia africana*, which falls under the *Bignoniaceae* species, can be found in South, Central and West Africa. It bears fruits that strikingly resemble cucumbers. This explains why the tree is also known as the cucumber or sausage tree in some regions in Africa. The tree has several uses, although it is important to note that the fruits are poisonous when unripe. The fruit contains a lot of fibres and seeds. The fruits can be baked and used to ferment beer or boiled and used as a red dye. The fruit and bark of the tree can also be ground, boiled and used as an enema to treat pediatric stomach ailments, particularly those caused by worms. Adults can also drink the boiled product to treat venereal diseases. In addition, the products can also be used to treat various kinds of skin ailments. Laboratory and chemical tests conducted in Nigeria have confirmed the tree's efficacy in inhibiting all microorganisms (Majoe, Online).

Almost all parts of the tree possess some spiritual and medicinal value. In some regions it is believed that hanging the fruits around a building can protect the building against virulent storms and hurricanes. It is also believed that the fruit can reverse infertility in humans. These claims have never been verified. However, there is no doubt that the tree possesses enormous verifiable pharmacological value. In this regard, the fruit can be processed into a powdery form that is used to treat ulcers. The unripe fruit is used to dress wounds, haemorrhoids and rheumatism. The tree also constitutes the primary source of medication for treating eczema and related skin ailments. Chemical analyses of different parts of the tree, including the roots, wood and leaves, have revealed that they contain *napthoquinones, dihydroisocoumarins, flavonoids* and *aldehydic iridoid derivatives*.

Tisanes

Aromatologist, Barbara Payne in her online article, 'Tisanes and their use for minor ailments,' describes several ailments that can be treated with this herb. She begins her piece by noting that the art of preparing tisane is an ancient one. One of the many ways by which herbal and floral tisanes can be prepared is by using proprietary products, which are usually in the form of tea bags. Its applications in Africa are numerous and include treating eczema, upset stomachs, cuts, bruises, minor burns, and grazes. Also, tisane may be taken simply as a refreshing drink.

Neem

Julia Cornborough, a clinical aromatherapist, is intimately familiar with the medicinal qualities of Neem (*Azadirachta indica A. Juss*), a medium-to-large-sized tree, which is characterized by its short straight trunk. African traditional healers have used products from different parts of the tree to treat a variety of human diseases. Recently, modern biomedical and medical experts have been making efforts to incorporate these products into formal healing strategies. For instance, in Nigeria, a number of experiments have been conducted to determine the efficacy of leaves from the Neem tree, also known as the 'wonder tree,' to combat malaria (see e.g., Edeinya, 1970). The results of these experiments have been very encouraging and go a good way in validating African traditional healing practices that have employed Neem tree products since the pre-colonial era.

Cornborough has reported on a number of other efforts on the part of modern medical experts to determine Neem's potency. Those involved in such efforts are unified in concluding that products from this tree are effective in treating several types of ailments, including but not limited to: fungi, cough, fever, diabetes, muscular pains, and some heart-related ailments. It has also been known to cure other ailments such as high blood pressure. Almost all parts of the tree, including the leaves, flowers, fruits, bark and seeds and the oil from its seeds are useful. Cornborough (Online) cites a number of laboratory studies (e.g., Khan and Wassilew, 1987), suggesting Neem's toxicity to cultures of fourteen common fungi, including members of the following genera: 1) *Trichophyton* – a type of fungus that infects hair, skin, and nails; 2) *Epidermophyton* – a 'ringworm' that attacks both skin and nails of the feet; 3) *Microsporum* – another type of 'ringworm' that attacks mostly the skin, but occasionally, the nails; 4) *Trichosporon* – an intestinal fungus; 5) *Geotrichum* – a yeast-like fungus that is known to attack the bronchi, lungs, and mucous membranes; and 6) *Candida* – another yeast-like fungus that may cause lesions in the mouth, vagina, skin, hands and lungs. Cornborough (Online) lends further credence to Neem's potency by recounting the fact that she has employed Neem products in her practice to successfully treat teeth and gum ailments such as tooth decay, bleeding and sore gums, and mouth ulcers. She further reports that, 'Neem has been found to reduce insulin requirements by up to 50% for diabetics, without altering blood glucose levels,' and alludes to recent studies that have shown Neem treatment to be effective in lowering high cholesterol levels (Cornborough, Online).

African Cucumis

Leaves, fruits and roots of the wild African *Cucumis* have been used to treat liver ailments, boost the body's immune system and as an anti-carcinogenic since pre-colonial times. Modern Scientific Studies have revealed that the leaves, fruits and roots of this ancient African plant contain *Cucurbitacins*, which has been described as a very bitter principle (Herbal Africa, Online). The studies further show that the plant can serve as a source of amorphous, bitter-tasting, crude toxin. African traditional doctors have used the African *Cucumis*, which possesses properties identical to the *Pedicellus melo* (bearing the Chinese name, *Tian Gua Di*), to remove lumps, eliminate fluid and also to relieve jaundice, treat acute and chronic viral *hepalocirrhosis*, liver cancer, persistent dyspepsia and epilepsy due to *windphegm*.

The plant or products from the plant are used as food in Tanzania. In Zimbabwe, particularly amongst the Bemba of the northern part of the country, the fruits of this plant are eaten raw or cooked as a vegetable. The Zulu of South Africa have historically administered a warm water infusion of the *Cucumis africanus* fruit pulp as an enema for relieving lumbago. The root of *Cucumis* can be soaked in cold water and used as medication for gonorrhea. When heat is applied to the same remedy, its properties change and transform it into an agent for inducing diarrhea in constipation patients. In West Africa, the seed of *Cucumis africanus* fruits are used as a remedy for roundworm.

African Ginger

This is a deciduous plant with large hairless leaves. Traditional healers in Africa have used ginger products for centuries to treat a number of ailments, including but not limited to, headaches, influenza, mild asthma, sinusitis and throat infections, menstrual cramps, mood swings, and hysteria, colds, flu and coughs. Ginger products have also been used as a mild sedative. Traditional Africans recognized ginger as an excellent remedy for indigestion, nausea, gas, colic and congestion. Furthermore, they recognized and used ginger as a circulatory stimulant, an important remedy for fevers, as well as some cardiovascular ailments.

Modern scientific studies have confirmed the medicinal properties of ginger (Herbal Africa, Online). There is evidence that ginger possesses properties capable of reducing cholesterol levels. It acts as an anti-coagulant agent, a vascular stimulant and body cleanser. It also relieves motion and morning sicknesses. Furthermore, it helps to remove toxins through the skin and through increased kidney filtration. The herb also relieves vomiting to sooth the stomach and spleen. It is useful as a supplement for heartburn and *halitosis* (i.e., 'bad breath').

Artemisia ('Wormwood')

The African herbal plant, *artemisia*, was named after the Greek goddess, Artemis. The plant has been used in Africa since the pre-colonial era. *Artemisia afra* is a common species of this plant that happens to be indigenous to Africa. This explains the surfix, 'afra.' Ancient Egyptians used the *Artemisia* for medicinal and religious purposes centuries ago. By some account (e.g., Herbal Africa, On Line), *artemisia* is the oldest and best known medicinal plant in South Africa in particular and Africa in general. The plant's stems, roots, and leaves have historically been used to treat coughs, colds, fever, loss of appetite, colic, headache, hypertension, intestinal worms, and malaria. It is taken as an enema, poultices, infusions, body washes, lotions, smoked, snuffed or drunk as a tea.

Gladiolus

Initially, some Europeans had erroneously taken this ancient African plant to be a similarly-looking plant found in Holland. Thus, European travelers in the Kwa Zulu Natal region in the 1800s labeled the plant as *G. natalensis*. However, *Gladiolus* is indigenous to sub-Saharan Africa and can also be found in Arabia and Madagascar. Products from different parts of the *Gladiolus* plant have been used to treat dysentery, constipation and diarrhea, lumbago, headaches and rheumatic pains.

Helichrysum

This plant has over 600 different species, of which 245 are found in the southern Africa region alone. It is a very potent plant with a lot of medicinal use. Ancient Africans recognized the plant's potency several centuries ago. Today, modern scientists confirm that the plant possesses more anti-inflammatory qualities than the German chamomile and contains more tissue regenerating capabilities than lavender. Also, it is more capable of fighting the formation of scar tissue than Frankincense. The oil from *Helichrysum* has been found by European researchers to generate tissue, reduce tissue pain, helps improve skin conditions, circulatory function, prevent phlebitis, help regulate cholesterol, stimulate liver cell function, and reduce scarring and discoloration. It is an anti-coagulant, anti-catarrhal mucolytic, expectorant, and antispasmodic.' It has also been known to fight certain types of hearing loss.

Hibiscus

Ancient Africans used the hibiscus to treat several ailments. Presently the hibiscus plant continues to serve a variety of medicinal purposes. The Guineans of West Africa use hibiscus products as mucilaginous and a tonic for curing heart and stomach ailments. They use hibiscus seeds, which have a musk-like odour as a perfume. Senegalese have used the same seeds as a remedy for eye diseases and dysentery. Modern scientific research has revealed that hibiscus seed oil contains 55.9% of stearic acid, 9.1% of palmitic, 32.6% of oleic and 2.4% of linoleic (Herbal Africa, Online). The root of hibiscus is chewed to relieve heartburn. The plant is also used to treat headaches and colic. It is also effective in treating 'bad dreams and help prepare men for courting.' Furthermore, it has been used to treat penile irritation, vaginal infections, venereal sores, and *urethritis*.

Colonialism, Christianity and traditional medicine

It is incontestable that traditional African medicine has regressed tremendously over the years. Most people familiar with the traditional healing scene in Africa have at least a story or two to tell about a frustrating experience with a traditional healer who was unable to provide a badly needed healthcare service but refused to acknowledge his/her limitations. The inability of a traditional healer in Africa to deliver a health service today may be rooted in one of two factors. It may be either an upshot of incompetence on the part of the 'healer' or a product of the fact that the 'healer' is an impostor. The latter is a growing and nagging problem perpetuated by mendacious and dishonest individuals who are more interested in the patient's money than with her well-being.

However, the former, that is the problem of incompetence, is one that is rooted in a number of socio-political developments that have occurred in the history of Africa. Arguably the most prominent of these developments are colonialism and the introduction of Christianity on the continent.

I have already drawn attention to the impact of colonialism on African tradition and culture in different domains. In the area of medicine and public health, the record of colonialism is inconsistent and remains a subject of heated debate amongst health historians and medical anthropologists. According to some, there is no evidence that colonial governments sought to decisively destroy African traditional medicine and healing practices (e.g. Feierman, 1985). In practice, colonial governments, which more often than not, operated on a shoe-string budget, were usually incapable and unwilling to provide modern healthcare services to members of the indigenous population or the 'natives' as they were known in colonial jargon. Consequently, they had no choice but to tolerate traditional healing practices. It is important to here state that 'tolerating' did not mean encouraging. In fact, colonialism clashed violently with traditional medicine in the area of political authority. As stated at the outset of this chapter, traditional healers were at once, priests, magicians, spiritual and political leaders. This did not bode well with colonial authorities, especially in colonies such as the Belgian Congo, where direct rule prevailed.

In general, as Kent Maynard (2004) noted, colonialism strived to break the links between healing and public authority, thereby effectively wresting control over economic production away from traditional healing systems and cognate indigenous institutions. In territories that were under indirect colonial rule, such as the British Southern Cameroons (present-day Anglophone Cameroon), traditional healing systems benefited from a measured degree of colonial government support (Maynard, 2004). In these territories, colonial governments sought to break only those ties between traditional healing and public authority that posed a direct and significant threat to colonial rule while acquiescing and sometimes, directly buttressing healing institutions with political import deemed crucial to the success of indirect rule. Kent Maynard's work on the 'history of public health and well-being in Cameroon' contends that Christian missions and their agents, including catechists, priests, pastors, missionaries and schools, were more active and more effective than colonial governments in destroying traditional healing systems.

Thus, it is safe to conclude that Christian missionaries were at the forefront of the war against traditional African healing practices. In this regard, they preached that African traditional healing practices and the concomitant rituals, which were inextricably intertwined with African religion, were antithetical to Christian doctrine. Christian missionaries were bent, in the first instance, on supplanting African traditional religious practices, which they deplored, with Christianity. Yet, religion and healing are inseparable in African tradition. The President of the Zimbabwe Traditional Healers Association, Gordon Chavunduka (Online) provides two reasons for this strong bond between African religion and traditional healing practices. First, traditional Africans have a relatively broad theory of illness, which includes theology. In this case, the theory does more than explain illness and disease. It also attempts to explain the relationship between God and the universe. Secondly, and as stated above, healers were invariably religious leaders in traditional Africa.

At least two questions deserve more than passing attention here. First, why were Christian missionaries bent on effacing African traditional healing and religion? Second, how successful were they in actually eliminating traditional healing practices? A partial response to the first question is implied above. Christian missionaries argued that traditional healing practices and the concomitant rituals are antithetical to Christian doctrine. Is this in fact true? Not according to the World Council of Churches (Chavunduka, Online). Christian missionaries falsely associated traditional African healing practices with witchcraft and also falsely accused traditional healers and anyone subscribing to their theories of healing as witch worshipers. These charges possessed no grain of truth. Traditional African religion neither worships nor encourages the belief in witchcraft or witches although it acknowledges their existence. Attempts to demonize African traditional healers for acknowledging the existence of witches were at best hypocritical. This is particularly because as attested to by relevant Biblical passages, Christians also recognize witches or what they call demons as a real phenomenon. Within the framework of traditional African healing and religious practices, God is seen as the Almighty and Alpha and Omega, who deals with the macro-picture, while ancestors are supposed to be concerned with micro-issues. In addition, Africans see ancestors as intermediaries through whom people pray to God.

The actual rationale for indefatigable efforts on the part of Christian missionaries to annihilate African traditional healing practices can be appreciated only within the broader context of the colonial enterprise. In this connection, it is important to note that Christianity and colonialism were in a *quid pro quo* relationship in which each complemented the other's efforts. Seen from this perspective, it is arguable that efforts to discourage traditional healing practices were part of a more elaborate strategy to eliminate any real or potential competition with Western medical and pharmaceutical interests. It takes little imagination to appreciate how the use of traditional healing methods and techniques can pose a viable threat to Western-trained health professionals as well as Western suppliers of pharmaceutical products, healthcare and medical equipment. African traditional healers use the same medicinal plants that are used to prepare so-called modern medications. Given the dwindling supply of these plants in the increasingly depleting forests of Africa, it is easy to appreciate this concern on the part of Western pharmaceutical industries.

Despite the efforts of early Christian missionaries, traditional healers remain extremely popular in Africa. The reason for this is simple. Traditional healers are more successful in treating some ailments than their Western-trained counterparts.

This revelation is hardly novel. In fact, as far back as the 1700s, European visitors to pre-colonial Africa had acknowledged the fact that some native treatments were more effective for some afflictions than European therapies. In this regard, Bosman (1705) noted that some Europeans were cured by traditional African remedies in a number of instances in which European physicians had given up hope.

The effectiveness of traditional African healing methods and techniques is not unconnected to the fact that some of them have withstood several centuries of testing and re-testing. To be sure, contrary to popular belief, these methods and techniques are at once non-scientific and scientific as well. Traditional healers obtained scientific medicines from natural plants. As stated above, most of the medicinal plants prescribed for certain specific ailments by traditional healers are deemed appropriate when put to contemporary scientific tests. Chavunduka (Online) elaborates on this process in the following words,

> This empirical knowledge has been developed through trial and error, experimentation and systematic observation over a long period of time. The major sources of non-scientific or subjective knowledge are the various spirits believed to play a part in health. The social and psychological methods of treatment developed from this unscientific base often bring good results.

Development implications of traditional medicine

The importance of this chapter can be appreciated at a number of different levels, including the following. First, it contributes to efforts seeking to reverse the damage that has been visited upon African traditional healthcare and healing methods by colonial authorities, Christian missionaries and other agents of Western civilization. These agents were exceedingly successful in tarnishing the image of African traditional medicines, healing methods and techniques. Second, it draws attention to the versatility of traditional African herbs and their promise for contemporary efforts to address the healthcare needs of Africans. Third, and in concert with the main theme of this book, it accentuates the importance of tradition and culture in the healthcare domain as an important determinant of the outcome of development efforts. Finally, the chapter underscores the importance of cultural knowledge as an indicator of healthcare quality.

Some elucidation of this last point is in order. The number of health professionals of non-African origin working in one capacity or another in Africa has increased significantly since the mid-1980s. This is when the need for health professionals on the continent began skyrocketing, thanks to the HIV/AIDS pandemic, civil unrests (in Sudan, Congo D.R., Sierra Leone, Liberia, Rwanda, Burundi) and outbreaks of previously unknown diseases (e.g. Ebola in Congo PR), and hunger and malnutrition (Niger, and Dalfur in Sudan). The effectiveness of these professionals is significantly compromised by their limited, or complete lack of, knowledge of African culture and tradition. Evidence suggests that healthcare professionals are less effective when they possess limited or no knowledge of their clients (Cole, 2004). Professionals with limited or no knowledge are wont to harbour biases fueled by stereotypes about their clients. Studies in the United States (e.g. Smedley, et al., 2003; Ferguson and Candib, 2002) have demonstrated that such stereotypes invariably lead healthcare professionals to deliver inferior healthcare services to non-Caucasian patients. It is therefore safe to conclude that this problem is likely to be magnified when such professionals are working in African settings where the patients are not only ethnically and racially different, but also linguistically and culturally distinct. Professionals in the healthcare policy field in Africa can significantly improve their effectiveness by paying attention to the unique characteristics of African customs, tradition and culture. Doing so entails a lot of deliberate effort on the part of these professionals whose analytic and diagnostic thinking is, as Cole (2004) has noted, so routine that they tend to ignore the European origin of their so-called objective scientific worldview.

Traditional African healing and healthcare methods tend to be inseparably bound to spirituality. Health professionals are increasingly recognizing that the way people perceive or interpret, and act on illness is a function of their culture. In essence, interpretations of the raw events of sickness are only meaningful to the extent that they are culturally constructed (Coreil et al., 2001). Similarly, the ecological model of health situates human behaviour squarely within a broadly defined physical, biological, and socio-cultural environment. Within this framework, decision-making relating to treatment choices, which may span the gamut from modern to traditional alternatives, is analyzed in terms of the influence of factors such as climate and seasonal conditions, subsistence patterns, social organization (including household dynamics) and ethnomedical systems (e.g., beliefs about illness and treatment options).

Although much is often made of belief systems, there are more similarities between Africa and the West in this regard. If these similarities are often not obvious, it is because Westerners are wont to deride African customs as primitive, superstitious and meaningless. Consider the question of spirituality, which is paramount to health and wellbeing in African traditional thought. The African traditional doctor believes he is able to succeed in his trade only to the extent permitted by supernatural powers over which he has no control. This is not different from modern medical experts who acknowledge the limits of science and have encountered several situations in which a remedy that registered positive results in one patient failed to do so, or actually registered negative results, in another.

The ecological model draws attention to the importance of the immediate environment. By extension, it acknowledges the potency of local or indigenous knowledge. Although early anthropological research treated indigenous customs in non-Western settings as barriers to successful innovations, the discussion in this chapter has revealed that these customs can actually facilitate innovations in the healthcare field. Contrary to popular opinion, many African traditional practices are in concert with, rather than antithetical to, health promotion efforts.

The tendency to denigrate African customs on the part of Westerners prevented the incorporation of indigenous knowledge, particularly mothers' knowledge and practices in the area of child health, in maternal education efforts during the colonial and immediate post-colonial era on the continent. Africans had recognized the importance of breastfeeding to children from time immemorial. However, as part of modernization efforts in the late-1950s and 1960s, Western propaganda campaigns aggressively promoted bottle-feeding as an alternative to breastfeeding. Consequently, those with the means adopted bottle feeding in lieu of breastfeeding. A few years later, and subsequent to scientific research demonstrating that breastfeeding was required for the normal development of infants, international health authorities reversed themselves and began advocating breastfeeding, but not before much irreversible damage had been done.

Research on the cultural construction of disease has proven to be very instructive. For instance, as Coreil and associates (2001) noted, research in non-Western settings has revealed etiologic beliefs related to different types of ailments. Insights from such studies have been instrumental in highlighting the importance of explicating culturally valid definitions of terminologies that are typically employed in reference to different ailments.

That some attention is being paid to the cultural context of health is a welcome development. However, much remains to be done. Western and Western-trained health professionals will do well to acknowledge traditional healing practices regardless of the extent to which they understand such practices. Individuals with a strong adherence to spirituality may consider the violation of a spiritual taboo to be more harmful than death (Cole, 2004). Traditional Africans believe that the Almighty God is the only one who can create and take away life. Thus, mention by any mere mortal, including physicians, of impending death to a patient is unlikely to be taken kindly by the patient. Such mention may be seen as an invitation to death or as a wish of ill luck.

There has been much progress in science and technology in the recent past. Much of this progress has occurred in the area of bio-medical sciences. At the same time, many new diseases have emerged while some old ones have resurfaced. The impact of this development has been most devastating on Africans. Witness for instance the problem of HIV/AIDS, which has reached pandemic levels throughout the sub-Saharan region. Coupled with chronic and nagging problems, including but not limited to poverty, malnutrition and opportunistic diseases, such as tuberculosis, this pandemic threatens to socio-economically cripple the entire region. Successfully combating this problem will require the input of not only the orthodox health and biomedical field but also that of entities within the non-conventional scientific community. Prominent amongst the actors in this community are African traditional healers. Despite the major role that traditional healers and medicine have always played, they have been largely ignored in the healthcare delivery systems of African countries. Failure to recognize the role of traditional healers has meant that most resources have been channeled exclusively to conventional programmes. Yet, with the ever-rising costs of sophisticated and prohibitively costly technologies, the healthcare needs of the poor will continue be ignored. It is therefore, necessary to promote the exchange of information between Western biomedical experts and traditional

African healers. As noted earlier, Western scientists are generally skeptical about traditional African medical practices. Therefore, there is a need to acquaint Western biomedical and cognate scientists with traditional healing methods, techniques and medicines.

Such efforts are already ongoing. In this regard, the National Center for Complementary and Alternative Medicine (NCCAM) of the National Institutes of Health (NIH) held a conference under the theme 'African Healing Wisdom: from Tradition to Current Applications and Research,' in Washington, D.C. from 6-9 July 2005. The purpose of the conference was 'to explore the uniqueness, wealth and complexity of African traditional medicines, as well as the potential role they could play in addressing some of the crucial health challenges of our era' (NCCAM, Online). Exchanges between African traditional healers and Western biomedical scientists hold much promise for global healthcare efforts. As the discussion in this chapter has demonstrated, health-promoting behaviour is not novel in Africa. Despite the rush to inject so-called new ideas and novel ways to deal with life situations, 'new' is not necessarily synonymous with 'better.' As biomedical scientists and other health professionals ferret for strategies to deal with the intricate health problems of contemporary Africa, particularly the HIV/AIDS pandemic, it may not be a bad idea to re-examine the myriad herbs that have been employed by traditional healers from ancient times.

Chapter 10

Traditional Architecture
and Housing

Introduction

More than half a century ago, G. Anthony Atkinson, the Colonial Liaison Officer at the Building Research Station of the Department of Scientific and Industrial Research in British Tropical Africa, noted one deficiency in the literature on life in Africa. The deficiency had to do with the lack of attention to African housing strategies. The absence of housing in this literature, Atkinson (1950: 228) remarked, 'is astonishing when one considers that shelter is one of man's basic needs, and its provision, through the art of building, one of his most important activities.' He went on to argue that,

> Not only does the house form the physical background against which a community develops but, in the form of the shrine and temple, affects its mystical life. The way in which the different peoples of Africa have developed house forms suited to their customs of living, to the building materials close at hand, and to the local climate, is a fascinating subject (Atkinson, 1950: 228).

Since Atkinson lamented the tendency to ignore African housing, architecture and spatial organization strategies in the relevant literature, a few works attempting to address this shortcoming have appeared (see e.g., Ojo, 1966; Lebeuf, 1967; Oliver, 1971; Prussin, 1974, 1999; Denyer, 1978; Schwerdtfeger, 1982; Falade, 1990). However, these works tend to be generally weak on two important fronts. First, they are generally descriptive. Second, they fail to determine the implications of traditional and customary practices for development efforts in Africa. Thus, many gaps exist with respect to our knowledge of the implications of traditional and customary practices in the housing sector on development. This chapter seeks, amongst other things, to contribute to efforts aimed at bridging these gaps. I begin by painting a vivid picture of African traditional architecture, building techniques, housing strategies, and spatial structures. In the process, I discuss the colonial policies and other measures that sought to replace customary practices in the human settlement development domain with Western or so-called modern varieties. Finally, and before concluding the chapter, I identify and discuss the implications for contemporary development initiatives in Africa of efforts to supplant African traditional and customary building and environmental design techniques and strategies with modern substitutes.

Traditional architecture and building techniques

Western scholarship has never paid any significant attention to African traditional architecture. The conventional wisdom in the Western world holds that Africa, perhaps with the exception of Egypt, has no history of accomplishments in the area of architecture. In fact, Labelle Prussin (1974: 183) narrated a story that goes a good way in buttressing this assertion. Her tale relates to a team of photographers from a well-known American news magazine, who were dispatched to Africa to photograph architecture for a picture story on world architectural styles. After spending some time in Africa, the photographers returned to the United States without any photographs. When asked why they returned empty-handed, the photographers stated without equivocation that there was nothing of architectural value throughout the continent of Africa. In the photographers' own words (quoted in Prussin, 1974: 183), 'All we could find were a bunch of mud huts.' Somehow, Prussin believes – and I tend to disagree – that the photographers' 'comments reflect a generally prevalent attitude that monumentality and permanence are prerequisites to architectural definition and that vernacular architecture is lacking in both identity and meaning' (Prussin, 1974, footnote 2, p. 183). Pre-colonial Africans had their own share of monumental structures. In his thoroughly researched and well-written book on *African Cities and Towns before the European Conquest*, Richard Hull (1976: 50) draws attention to such monumental structures when he talks of the 'massive reed palaces in the capitals of Buganda and Bunyoro in East Africa.' He then goes on to characterize the architectural form of some of the buildings comprising the palaces as conical and often rising to a height of more than twenty feet with a diameter at the base of thirty-feet. One of the photographs in Henry M. Stanley's 1876 work, *Through the Dark Continent* (Vol. 1), (Harper Brothers), is that of a Serombo King's palace, which stood at thirty feet and had a base diameter of fifty-four feet. The Monarch's Obelisks at Aksum in present-day Ethiopia, which I discuss in more detail below, attained heights of more than one hundred feet. Other monumental structures that were present in pre-colonial Africa were the Kasbahs of Morocco. Kasbahs rose to heights of as many as ten storeys.

The photographers' misrepresentation of the reality of indigenous African architecture could have been considered an honest error were it not for the fact that their remarks dovetail neatly into the broader agenda of Westerners to devalue the accomplishments of pre-colonial Africans in every conceivable domain. This agenda, as noted earlier on in this book, dates back to the period when Europeans first made contact with Africa. The fabrication of the myth of Africa as a 'dark continent,' which was later employed as a rationale for invading and colonizing the continent, was part of this agenda.

The role of early explorers in devaluing the accomplishments of Africans in the architectural domains is evident in the following tale recounted by Labelle Prussin (1974). Three explorers, each recounting his experiences in different regions of Africa, all employed a similar picture purporting to depict traditional African housing. This is despite the fact that each of the three regions in question had its own unique housing in terms of architectural style and nature of building materials. Such accounts were, and remain, typical. The stereotypical image of Africa in the West leads Westerners to believe that for any given entity 'once you have seen one, you have seen all.' Yet, nothing could be further from the truth. As demonstrated below, several architectural styles are indigenous to Africa. These styles reflect the geographical differences characterizing the different major regions of the continent. Thus, one encounters different architectural styles as one moves from Cape Town in the southern tip of the continent to Cairo in the northern extreme, and from Dakar in the extreme western part to Mogadishu in the western end of the continent.

African architecture cannot be fully appreciated unless it is seen as part of the artistic expression of African people as opposed to nothing more than an aspect of the shelter production process. Also, it is invariably more useful to deviate from the orthodoxy that confines the meaning of architecture to gigantic and permanent structures based on detailed blueprints. Doing so permits us to entertain as part of African architecture artistic expressions such as the finely carved wooden pillars and columns that were used throughout most of coastal West Africa, the circular clay domes that formed the magnificent conically-shaped roofs of buildings in the Sahelien region, the decorative metal pieces and/or bamboo that were used in the construction of doorposts, window frames, and roofs throughout the continent.

The artistic prowess of traditional Africans was given architectural expression in many physical structures that constituted an important feature of the landscape of pre-colonial Africa. One of the best known of these structures was the Ethiopian Monarch's Obelisk at Aksum, which dates back to 300 B.C. and 300 A.D. Here, Africans displayed their ingenuity in crafting what came to be known as freestanding masonry structures. The Obelisk, as mentioned above, stood some 100 feet tall and had false doors, windows, columns, and wall treatments complete with horizontal wooden reinforcements and outward projecting timber stumps. Despite the disparaging remarks of European visitors to the continent, there is a preponderance of evidence suggesting that Europeans were flabbergasted by the tremendous strides Africans had made in different domains. Europeans appeared particularly taken by the sophisticated artistic skills and construction engineering acumen of Africans. The fact that some European authorities worked fervently to confiscate and transport to their home countries African works of art lends credence to this assertion. Possibly one of the best-known efforts in this connection occurred in 1937 when Musolini ordered the looting and onward transfer to Italy of one of the Obelisks from Ethiopia. This engineering and artistic artifact, which stands some 25 feet tall, continues to be one of the most impressive and conspicuous features of the landscape of Italy to date.

By Labelle Prussin's account, the pilfering and pillaging of African art by Europeans occurred at a rate and magnitude few are ready to admit today. She cites the case of France, which controlled most of tropical Africa, particularly the sculpture-producing regions, and effectively became the main source of the many pieces of African art that made their way into museums throughout Europe in the 1900s. According to Prussin (1974: 184),

Increasing numbers of 'artefacts' and curios pilfered and pillaged during the decades of colonial expansion appeared in European museums and bistros, inspiring Picasso, Modigliani, and others. But, while one might carry off sculpture and decorative art for display to the Western world, architectural elements are more difficult to transport.

Their cumbersome nature notwithstanding, as the looting of Ethiopian obelisks cited earlier suggests, architectural artifacts were just as likely to be stolen and transplanted to Europe as items of a more portable genre. Prussin (1974) made this point by drawing attention to the fact that despite their weight, Egyptian architectural artifacts such as the obelisks, carved wooden columns, plaques in wood, decorative roof pinnacles, were severed from their sites and transported to Europe.

Although colonial government policies discouraged indigenous African architectural designs and construction techniques, colonial authorities acknowledged the merits of African accomplishments in these domains. Testament to this acknowledgment resides in the fact that colonial authorities were quick to adopt the traditional African techniques of using louvered doors and screened verandahs, as well as raising building structures a few centimeters from the ground as a means of facilitating ventilation and ensuring climatic comfort within buildings. As Prussin (1974) noted, French and British colonial authorities adopted these techniques in constructing colonial government buildings, especially the residential units for colonial government officials. Some of these structures exist to date in the former British and French colonies especially in West Africa.

Classification of African traditional architecture

Few regions in the world lay claim to as much variation in architectural style as Africa, particularly the sub-Saharan region. I employ the term, style, interchangeably with the term form. However, I hasten to draw attention to the more specific meaning of the term, which has to do with building genres of the past as classified by architectural historians. Thus, an architectural style can be defined as 'a definite type of architecture distinguished by special characteristics of structure and ornament' (Crisman, Online). As Phoebe Crisman explains in her piece in the *Whole Building Design Guide*, building '. . . styles developed at specific moments in time, in particular geographic locations, and within unique cultural conditions' (Crisman, Online).

While the northern portion of the continent is dominated by Arabic architecture, the sub-Saharan region boasts a considerably wide variety of architectural styles. Richard Hull (1976) identified several forms of building designs that were dominant in sub-Saharan Africa during the pre-colonial era. Prominent amongst these are the following: beehive or domical, bread-loaf-shaped, onion-textured, inverted cone, bullet-tip or egg-shaped, bell-shaped, rectangular with hump roof, square-box with pyramidal roof, ant-hill-shaped, box-on-dome, cone on cyclinder, cone on poles and clay cylinder, rectangular with gable roof, rectangular with rounded roof sloping at the ends, square, quadrangular or square with dome or flat roof, quadrangular surrounding an open courtyard, and cone on ground.

As equally diverse as the architectural forms are the construction materials. The diversity of these materials is a reflection of the wide range of physical, technological, socio-cultural, political and cultural environments comprising the continent. In discussing how geography and other natural features have impacted architectural form and choice of building material in Africa, Labelle Prussin (1974: 185) stated that,

> The climate of humid coastal rain forest belt, where there is little temperature change between day and night or even between wet ad dry seasons, calls for a shelter with a maximum of cross ventilation to ensure bodily comfort.

Gebremedhin (1971, citing Cipriani, 1938) has classified the type of houses in Africa, with particular emphasis on Ethiopia, based on the structural nature of the building materials employed. Resulting from this are three distinct house-types, including those built of: 1) rigid elements, 2) flexible elements 'planted' to the ground at one end, and 3) flexible elements 'planted' at both ends. Alternatively, African houses can be classified based on the processes involved in their production (Gebremedhin, 1971). From this perspective, and with respect to the Ethiopian housing sector, houses can be grouped into four main categories, including:

- those produced by piling building elements such as blocks, bricks or stones;
- houses produced by twining or tying building elements such as twigs, grass, cloth or leather, together with ropes; and
- houses produced by weaving together building materials such as bamboo and grass.

In terms of shape and form, African houses may be (Gebremedhin, 1971):

- Round with a conical roof;
- Rectangular with a flat roof; and
- Paraspherical.

The production of housing through piling tends to be slow and expensive. Two major reasons account for this phenomenon. The first is that piling usually entails the use of solid, and quite often, heavy building materials.

The second reason is that the foundations for such buildings typically require not only enormous financial resources but also a lot of time. The housing units resulting from twining or tying and weaving usually have no corners. Rather, they typically have cylindrical walls or hemi-spherical and conical roofs. The reason for this is simple. It is easy to produce plane surfaces by weaving or tying. Gebremedhin (1971: 110) states this more elegantly as follows: 'tying and weaving lend themselves admirably to the production of plane surfaces that have no abrupt change.' In explaining why woven and twined traditional African housing units tend to be windowless, Gebremedhin (1971: 110) stated thus, 'surfaces produced by the tying or weaving process do not accommodate openings easily and naturally.' Another well-known typology for African traditional architecture includes the following categories: tents and other temporary structures, Sudanese, impluvium, hill, beehive, ghorfas, kasbahs, and underground (Denyer, 1978; Prussin, 1974).

Tent and other temporary structures

The tent is the structure of choice for nomadic groups, hunters and gatherers and pastoralists throughout the continent. Examples of nomadic groups include but are not limited to the Tuaregs of the Sahel region (Niger, Burkina Faso, Mali), the Fulani's of most coastal West African countries (Cameroon, Nigeria, Benin, Sierra Leone, etc). These groups are interested in meeting their housing needs with the use of materials and structures that can be dismantled with relative ease. For some, an additional requirement is that the materials have to be light enough to permit easy transportation on the back of animals (camel and/or donkey). The Tuaregs exemplify this category. The design of the Tuareg's housing unit is relatively simple and usually comprises little more than a framework of hoops covered with mats. The entire structure can be dismantled, packed along with all of the belongings of a whole Tuareg family and transported on the back of a camel or donkey to a new location in short order.

Another group whose members use tents to satisfy their housing needs is the Tekna of Morocco (Andrews, 1971). Rather than a single tribe, the Tekna comprise several tribes that call a region in the south-western part of Morocco home. Some of these tribes, such as the Ait Üsa and Ait Lahsen are fully nomadic, while others such as the Lansas can be more accurately described as semi-nomadic in that they seldom move more than twenty kilometers from their inland villages. In 'Tents of the Tekna, South-West Morocco,' Peter Andrews (1971) gives a copious description of a typical Tekna tent. The superstructure of the tent consists of two main poles, which are set far apart at the base but converge at the peak to roughly form an equilateral triangle. The poles are straight and round, with a diameter of 5.5 to 6 cm at the widest part near the base. The superstructure is covered either with cloth or hide. In its complete form, the tent is provided with a ridge piece in which the tips of the poles lodge. The tent is always provided with fastenings for four guy ropes on each of the short sides. The pegs are either of wood or of iron.

The Massai of Northern Tanzania and southern Kenya, exemplify the pastoralists. In their indigenous setting, they are extremely mobile. Their architecture reflects this aspect of their lifestyle, as their housing units are typically portable. The units are usually made up of wooden frames, which can be transported by animals. The Fulani are also a nomadic people. However, their housing units are not necessarily mobile. Rather, the Fulani tend to construct their housing units of materials that are available on-site wherever they decide to establish their next home. The Fulani prefer hilltops. Thus, they typically move from one hilltop to another depending on the availability of fresh grass for their cattle. Their housing units are usually of clay walls, a circular floor plan and a conical grass roof. The units generally have as the only opening a very narrow and short door, measuring about 0.70m by 1.30m.

Temporary or semi-temporary housing units were also common amongst the natives of the grasslands of pre-colonial South Africa (Hull, 1976). These structures were typically lower, hemispherical and possessed an architectural form that strikingly resembled a beehive. The Pygmies of Congo also constructed similar residential units. It is believed, although no concrete evidence supports this theory, that the natives of South African grasslands borrowed this residential design from the Congo Pygmies. The Xhosa, Sotho and Tembu also construct identical units. However, the buildings of these latter groups tend to be more rounded and have 'a circle of bent saplings joined at the apex and strengthened by a series of concentric hoops, parallel to the ground' (Hull, 1976: 50–51). A number of other nomadic and semi-nomadic groups, including the Fipa of Tanzania and the early Kanembu of the Lake Chad region, also construct residential units that are similar to those afore-described. The units are typically made of straw that is loosely bound together. As noted earlier, a common thread running through the tents and other temporary structures discussed thus far is the fact that they require little time, energy and materials to build. Also, they can be easily dismantled.

Sudanese style

Sudanese style architecture consists of a variety of rectangular buildings developed around a common courtyard. Although some (e.g., Asojo, Online) believe that the origins of Sudanese architecture can be traced to the Arab world, I contend that its presence in Africa pre-dates the Arabic conquest and the concomitant introduction of Islam in the continent. My position is bolstered by the fact that this brand of architecture is as common in regions whose growth was influenced by interaction with Arabs as it is in those where such interaction was non-existent. In fact, as Abimbola Asojo (Online) noted, archaeological excavations at Nteresso in Northern Ghana have revealed remnants of buildings that possess characteristics of Sudanese architecture. Yet, this part of Ghana has no history of significant Arab influence. This renders hollow any theory positing that the roots of Sudanese architecture are located in Arabia or that Arabs introduced this brand of architecture in Africa. The major characteristics of Sudanese architecture include courtyard plans, flat or dome-shaped vaulted roofs of clay supported by palm fronds, parapet pierced with gutter pipes or channels, and mud brick walls. The fact that some of the first groups of Africans to convert to Islam, such as the Fulanis, Nupes and Khassonkes, were not known for building in Sudanese style also supports the theory that the style predated the advent of Islam in Africa (cf., Asojo, Online).

Sudanese style buildings typically assumed an architectural form that has been described as 'cone-on-cylinder' (Hull, 1976). The cone-on-cylinder structures are more permanent and have been historically constructed by sedentary agriculturalists and urban residents. These types of structures can be found in areas as far apart as Senegal and Cameroon, and in regions as geographically varied as the Savanahs of

Angola and the Democratic Republic of Congo, Tanzania, Malawi, Mozambique, Zimbabwe and South Africa. The units are typically constructed with or without a wooden superstructure. When they are constructed with a wooden superstructure, cylinders are made either with clay, and reinforced with twigs, or a series of wooden poles, which are then plastered over with clay. During the pre-colonial era, the roofs of such dwelling units were typically of grass, straw or thatch anchored to wooden or bamboo rafters. Alternatively, the units were/are constructed without any wooden reinforcement. In this case, the construction is usually by potters as opposed to masons. In pre-colonial West Africa, the same units were constructed from thatched cone reposed on a wholly clay or woven thatch cylinder.

Impluvium style

This usually comprised four buildings facing one another in a courtyard with gabled roofs. The totality of housing or other building units circumscribing a single courtyard are referred to as 'compounds' and usually belong to one extended family. Compounds are commonplace in West Africa. The royal palaces adhere to the impluvium model, if only on a larger scale. The palaces are considered sacred and constitute several courtyards, each of which has a designated function. The buildings surrounding the courtyard, especially in royal palaces, usually have roofs that repose on elaborately carved wooden columns. Again, as stated earlier, the palaces are remarkable for their magnanimous size. For instance, the palace of the king of the ancient Oyo Empire in Nigeria is said to have been more than twice the size of a modern sports stadium (Asojo, Online). The King's palace was usually a complex of elaborately designed rooms with sculpted walls, doors, columns and windows.

The pre-colonial cities of West Africa, the region with the longest history of urbanization, dating back centuries before the advent of colonialism in Africa, were generally made up of several compounds. Each compound consisted of a number of houses built around open courtyards of varying sizes. The courtyards contained pots to collect water from the roofs. The ancient cities, such as the city-state of Benin in Nigeria, were typically circular in shape and enclosed within a defensive wall. The similarity between this design and that of ancient Greek and Roman cities is unmistakable. Yet, one would be on *terra firma* to state that ancient African city builders were simply responding to a local need – that of protecting the city against any external threat – as opposed to imitating the Greeks or Romans about whom they knew nothing. The circular form of ancient African cities and their enclosure within a solid wall gave them a unique structure – one that served to emphasize the notion of 'placeness' and cement the bond amongst otherwise disparate units into an interlocking whole. Most importantly, the compound design served to articulate intimacy and invariably to strengthen extended family ties.

Hill style

This architectural style, as the name implies, was unique to hilly terrains. Africa is known for its numerous rugged terrains. This architectural style possesses a number of major features, the most peculiar of which are stone terracing, and their curvilinear and slender structures. The stone terracing provided a camouflage for the buildings against the background of natural rocks and served as a defensive mechanism. The significance of locating residential units on hilly or mountainous terrain requires further elaboration. Hills and mountains play an important role in African cosmology. This explains their prominence in several African iconographic and architectural contexts. Jomo Kenyata's *Facing Mount Kenya* (1938) is perhaps the most celebrated tribute to the ceremonial and spiritual importance of hills and mountains in African tradition.

Beehive style

Residential units assuming the form of a beehive were commonplace in the regions occupied by present-day South Africa, Namibia, Lesotho and Swaziland. Building units adhering to the beehive form had a dome-like shape. However, unlike units with domical roofs, the beehive was a complete building unit, which in its entirety assumed the shape of a dome. The super structure of beehive units were usually of wood covered with grass weaved tightly into the wooden super structure.

Ghorfas

These were typically multi-storey structures built of stone, sun- or furnace-baked bricks. As opposed to other structures discussed thus far, ghorfas are not used for residential purposes. Rather, they are barrel-vaulted chambers designed to store food products, particularly grains such as millet, corn and beans.

Kasbahs

These types of structures are important for several reasons. For the purpose of this book their importance is amplified by the fact that they serve to render hollow any theory to the effect that traditional African architecture is devoid of characteristics such as 'monumentality' and 'permanence,' for kasbahs were structures that usually attained heights of ten or more storeys. These structures were typically designed to house individuals with common ancestry. They are said to have originated as forts and adhered to an architectural style specific to Morocco (Asojo, Online).

Underground structures

As their name intimates, underground structures are typically located below ground level. These structures assume one of two common shapes, rectangular or circular.

Their roofs typically repose on wooden rafters. Earth is used to cover the roof as a means of camouflaging the residential unit as a whole. Like the kasbahs discussed above, underground residential structures also evolved to address defensive necessities.

Anthropomorphism in traditional African architecture

Anthropomorphism, the practice of attributing human characteristics or behaviour to animals or inanimate objects is of great antiquity in Africa. In relation to the built environment, Africans have been conceptualizing, visualizing and creating architectural space in terms that are analogous to those employed in characterizing the human anatomy for centuries before the arrival of Europeans on the continent. In contrast, some of the earliest and best-known evidence of anthropomorphism in architecture in the Western world date back only to the 16th century. Suzanne Blier uncovered this evidence in a letter from Michelangelo Buonarroti, in which she quotes him as stating as follows.

There is no question but that architectural members reflect the members of the Man and that those who do not know the human body cannot be good architects (Blier, 1983: 371). Since then, there has been a smattering of discussions on the subject, which has never attracted as much scholarship as one would expect given its importance. Within the folds of the thin literature on anthropomorphism in Western architecture can be found the following: Geoffrey Scott's *The Architecture of Humanism* (1999, Le Corbusier's *Vers Une Architecture* (1926), Steen Rasmussen's *Experiencing Architecture* (1962), Kent Bloomer's and Charles Moore's *Body, Memory and Architecture* (1977), and Roger Scruton's *The Aesthetics of Architecture* (e.g., 1979). More recent works on the subject include: M. Frascari's *Monsters of Architecture: Anthropomorphism in Architectural Theory* (1991) and Drake Scott's article, 'Anatomy and Anthropomorphism: Architecture and the Origins of Science' in the *Edinburgh Architecture Review* (see Scott, 2000). This recent flurry of works notwithstanding, anthropomorphism has never constituted a significant part of the structure and symbolism of how Westerners view elements of the built environment, especially houses.

In fact, the practice of attributing human characteristics to houses and other buildings is confined to theorists and practitioners in architecture and cognate fields. In contrast, as Suzanne Blier's study of the Tamberma (Togo) (1983) revealed, some traditional Africans have always thought of, and treated houses like 'humans beings,' and 'like humans, each house is said to be made from flesh, bones, and blood' (Blier, 1983: 371). The flesh, bones and blood in this case signify the clay, aggregate and water of which housing units are built.

The Tamberma, a Voltaic people located in Togo and Benin, West Africa, also suggest that the house, like a human being, may be female or male, and has a head, legs, hands, eyes, genitals, lips, tongue, nose, ear, stomach, and so on. They also believe that the house production process is analogous to that of conceiving and delivering a baby. Thus, the man who takes charge of the design and construction, and the woman who has the responsibility of plastering and decorating, the house are viewed as partners in art. Their roles are identical to those of the male and female ancestors whom the Tamberma believe, design and mold every baby from the sacred clay in its mother's womb. New houses are treated in the same manner as newborn babies. Furthermore, houses are expected to have a life span of about 56 years, the same as an average Tamberma person. Apart from referring to different parts of a house in the same terms employed in characterizing the human anatomy, these traditional Africans conceptualize dimensions of the house in terms of parts of the human anatomy. Thus, for instance, the builder's leg or arm span is used to determine the relative size of floor space, wall width or height, and so on.

The house, as a human being, may stand erect (vertical position) or lie down (horizontal position). On the one hand the house is said to be in a vertical position to an observer when the observer is on the same plain as the house. On the other hand, the building is considered to be lying down to an observer who is located at an altitude greater than that of the building. In technical jargon, the building's elevation is its vertical position while the plan is taken to be the building in its horizontal position. In its upright or vertical position, the house's eyes, that is, the windows, and its mouth, the door, are clearly visible. The house's eyes serve as organs of sight for its occupants. Apart from serving as the entrance into the house, like the human mouth, which is the entrance into the human body, the mouth of the Tamberma house is designed to receive greetings from the house's guests. African custom, even in societies with no tradition of anthropomorphism of architecture, dictates that a person's neighbours and acquaintances must stop by from time to time to greet his house by peering in and offering salutations to the mouth of the house even in that person's absence. It is believed that by so doing, each house is able to recognize 'its friends' and will therefore be willing to give them protective shelter should the need arise.

The traditional African considers the mouth of his house the easiest means to his house's innermost parts. Thus, when the African has something to share with the house, he offers it to the house's mouth. He may share his gourd of palm wine with his house by pouring some of the wine at the doorway. In this case, he pours a portion of the wine at the doorway in the same manner that he would in a friend's cup if he were sharing the drink with the friend. Elsewhere (Njoh, 1999), I have noted that the doorsill also plays an important function unrelated to the provisioning of shelter in the traditional African's life. As mentioned in Chapter Four, the approximately 45-cm-high doorsill of the traditional abode in Meta country, Cameroon serves to gauge the maturity of a child. Once a child is matured enough to scale the entrance threshold or doorsill, then she is deemed fit to wean off breast milk. Thus, the mouth of the house can be seen as playing a crucial role in family planning in traditional African society.

As one walks into the traditional house, one finds stones that are used for grinding grain, such as millet, corn and so on. Thus, the stones are called the house's teeth, particularly because they function as incisors that chew the grain into flour or other food products. The lintel at the top front portal is known as the tongue, while the extended rim protruding downward from the doorframe is known as the lip. A number of other important anatomical parts of the house include, the walls that connect the different small units making up the house. These are also called joints in the same manner as human joints. They play the same function, that is, connect different parts of the house as human joints do. The granaries are known as the house's stomach, for it is where the Tamberma store their food. The Tamberma build houses with flat roofs. Water from the roof is drained by means of a drainpipe that lets off the water to the side of the building. The drainpipe is called the penis, for it is believed that the house is urinating when it discharges rainwater that has accumulated on the roof through the drainpipe.

A traditional African woman, especially in a polygynous setting, does not share her room permanently with her husband. However, the husband may pay her conjugal visits by prior arrangement. When this happens, the man is said to figuratively into the wife's vagina and consequently her womb. Thus, the bedroom is called the womb of the house, for it is from here that babies originate. The bedroom also plays a critical ceremonial role in Tamberma society during initiation rites. Young initiates are usually required to assemble in the bedroom of the village founder's eldest wife, and it is here that the final step of the initiation rites is performed. Once this final step has been performed, the novices are then ordered to crawl out one-by-one, simulating multiple-delivery. This, amongst other things, is designed to impress upon the young initiates that they are brothers or sisters from the womb of the same village mother.

Building materials, architectural form and function

Foremost amongst the strengths of traditional African architecture and building techniques are: 1) the fact that they are a function of the immediate natural environment in terms of form and style; and 2) depend on locally available materials. Not only are the materials of the natural variety, they usually come at minimal cost. In most cases, as Gebremedhin (1971) has noted, they are usually free of charge, especially when no account is taken of the labour input necessary for their extraction and, in some cases, transportation. Another strong attribute of traditional African architecture is its compatibility with relevant natural conditions, especially climate, geography, topography and geology. To illuminate this particular attribute of African traditional architecture, I examine the architecture of West Africa with a view to showing how through many centuries of trial and error, inhabitants of the regions have succeeded in crafting house-building methods and techniques that maximize the utility of available materials as well as ensure the safety and comfort of the buildings' occupants.

The traditional architecture of West Africa, particularly the form of the houses and the choice of building materials, as Jean-Paul Bourdier and Trinh Minh-Ha (1985) noted, reflect the climate, soil and lifestyle of the people of the region. Thus, each

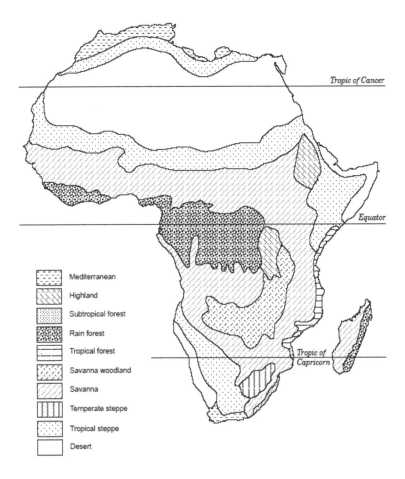

Figure 10.1 Map of Africa's vegetation
Source: http://www.fga.freac.fsu.edu/pdf/africa/africa_veg.pdf

architectural type corresponds to a specific geographic zone. The views of Bourdier and Minh-Ha on this issue mirror those of Labelle Prussin of a decade earlier. Prussin observed a logical fit between architecture and nature throughout Africa. In the case of West Africa, she noted that the architectural forms one observes in the region reflect its varying geography. Thus, as one moves along an imaginary longitudinal axis following the Greenwich meridian from Accra, Ghana, through Timbuctu, Mali, one traverses a series of horizontal climatic belts comprising humid rain forests near the coast, a derived woodland savannah inland, which gradually gives way to a grassland savannah and finally turning into a semiarid desert.

The inhabitants of these different geographic regions have demonstrated exceeding ingenuity in devising strategies to address their architectural needs. In this regard, Prussin (1974: 185) noted that 'the climate of the humid coastal rain forest belt, where there is little temperature change between day and night or even between wet and dry seasons, calls for a shelter with a maximum of cross ventilation to ensure bodily comfort.'

This explains the fact that the traditional buildings on the coast of West Africa typically boast some variant of louvered or natural opening. It also explains the proclivity towards bamboo walls that simulate openwork screens. This screens and the louvered openings have one major purpose, namely to encourage air circulation. This is also the function of the raised floors that typify traditional African coastal buildings. The floors, which are usually raised several centimeters off the ground, are designed to trap and circulate within the building, breezes from the ocean.

In the rain forest, the wind pattern is markedly different. Here, the winds tend to assume a cross pattern. Accordingly, the traditional African builder tends to adopt a rectangular form or design for his buildings. This form has proven more suited to the cross breezes characteristic of the rain forests. Further inland, the builders face a different situation, namely a savannah climate, which typically comprises a brief rainy season and a long dry season.

The extended dry season of the region also witnesses frequent and extended bouts of virulent harmattan winds originating in the Sahara Desert. In addition, daily temperatures in this region tend to swing rapidly between extremes. Temperature changes as high as $30°$ to $35°$ Fahrenheit are not unusual. To cope with these conditions, the inhabitants of this region designed buildings that are capable of simultaneously protecting them from excruciating winds and the concomitant cold temperatures and providing a cool respite from the baking heat produced by the midday intensive solar radiation for which the Sahara Desert is well-known. The walls of the traditional cylindrical or round housing structures of this region are thermostatic and have proven for centuries to be up to this task. In practice, these walls help to concentrate thermal radiation within the interior of the building. This is a function that rectangular buildings and non-clay buildings are incapable of performing. Also, the absence of windows and/or large openings on the cylindrical or round buildings, which some researchers have noted (see e.g., Gebremedhin, 1971) as a deficiency of such units, is by design and meant to maximize the thermal properties of the thick clay walls.

The utility of round buildings in the semi-desert and grassland regions is not limited to their ability to protect the inhabitants from the adverse climatic conditions of these regions. As Prussin noted, the softly rounded curvilinear surfaces of the cylindrical buildings paradoxically have the same visual effects as the tall trees of the forest zone. These surfaces, especially when combined with the rough textured clay walls, tend to 'eliminate the harsh, irritating contrast between light and dark created by perpendicular intersecting planes, and convert it to softly graded shade and shadow' (Prussin, 1971: 187). Rectangular shapes, which can only aggravate the problem afore-described, are not problematic in the forest areas, where the tall trees act as a light filter, subduing the sharp rays of the sun.

Development implications of non-indigenous architecture

As Hull (1976: 54) observed, 'popular magazines since the mid-nineteenth century have associated most traditional Africans with round, thatched-roof mud-walled huts.' Thus, according to popular opinion, all pre-colonial Africans lived in circular mud huts roofed with grass or thatch. This view is misleading given, as shown in this chapter, that a variety of architectural styles, including rectangular, quadrangular, and octagonal, and a wide range of building materials, such as stone, wood, and clay, are of great antiquity throughout the continent. I will elaborate on the strengths of traditional architecture and building materials in a moment. For now, lets examine in some detail the criticisms that have been leveled against traditional architectural practices by proponents of modern design and construction technology.

Proponents of modernity contend that traditional African building techniques and processes are, amongst other things,

- Oppressive to women;
- Impractical given the dwindling supply of local building materials;
- Unhealthy, particularly because they are poorly lit and inadequately ventilated;
- Susceptible to termite attack;
- Constantly in need of maintenance; and
- Aesthetically unappealing.

The charge that traditional African architecture and building processes are oppressive to womenfolk is at best based on anecdotal evidence from Kenya, where the customary norms and values of the Maasai place the burden of residential construction on women. To appreciate the implications of this for women, one must first understand that the gender-based division of labour already places a disproportionately large share of all the work in the domestic sphere on women. Furthermore, it is important to understand that the rapid expansion of human settlements has been increasingly encroaching on the natural environment thereby further elongating the distance to the source of local building materials. To the extent that this is true, it is easy to understand how placing the burden of residential building construction on women can easily translate into some form of oppression on womenfolk.

As stated earlier, the case of the Maasai is unique and must not be treated as the norm. In pre-colonial Africa, women and men participated in the residential development process. Thus, were the claim of a dwindling stock of local material true in the pre-colonial context, the responsibility for traveling long distances to acquire building materials would have fallen squarely on the shoulders of men and women alike. Even in contemporary times when men have left most domestic work to women, this claim overstates the problem. While it is true that urbanization is increasingly encroaching on the natural environment in Africa, only the few areas where local building materials consist of wood are witnessing a noticeable decrease in their stock of building materials. The areas in which local building materials consist primarily of clay – and as shown above, such regions are in the majority – have not been affected. Clay continues to be an abundant local building material throughout the continent. This fact renders porous any theory positing that the stock of local building materials in Africa is waning.

I concede that some traditional African residential units, particularly the cylindrical ones, are poorly lit and inadequately ventilated. However, it is easy to understand why traditional African residential structures in some regions are windowless. As noted earlier, there is a thermostatic rationale for minimizing the number of openings on buildings in regions such as the sahel, which have extreme temperature variations. As for the charge that traditional African buildings are constantly in need of maintenance, this is not supported by empirical evidence. Buildings of durable local materials such as clay and stones do not require more maintenance than those of so-called modern materials such as cement, glass and plywood. Critics of traditional building materials also tend to exaggerate the weakness of these materials when they claim that they are susceptible to termite attack. This charge assumes that wood constitutes the exclusive or dominant local building material in traditional Africa. This is certainly not true. There is a preponderance of evidence demonstrating that clay and stones are more commonplace than wood in the residential construction industry in traditional Africa.

The strength of traditional building materials is especially amplified when examined against the backdrop of the so-called modern alternatives that have been proposed by governments and Western change agents throughout Africa. Some of these governments forbid traditional building materials not on account of any real or perceived structural weakness on the part of the materials, but on grounds that they constitute a sign of backwardness. This mentality is at the root of efforts that often result in the hastily constructed, dreary tiny, and closely packed tin roofed housing units that have become a common feature of the outskirts of urban centres throughout the continent. These units literally become ovens under the intense heat of tropical Africa.

To be sure, African traditional building materials are not without blemish. However, despite the few noted drawbacks it is necessary to emphasize the fact that African architectural styles and construction materials have always been contextual, place- and time-specific. The importance of these three attributes of architectural style and construction components for development efforts on the continent cannot be overemphasized. The notion of context draws attention to a critical but often ignored dimension of housing in the development plans that have been drawn up and adopted in African countries since the colonial era. This dimension of housing has to do with the building unit's immediate physical, cultural, and economic environment. While pre-colonial Africans did well to relate to this context, colonial and post-colonial planners have failed to do so. In other words, pre-colonial Africans were 'contextural' in that they fully understood the context within which they were functioning and developed clear strategies with respect to how to anchor their dwelling units to the geographic, economic and cultural conditions that prevailed at the time. To be sure, this context within which pre-colonial Africans functioned was never static. Instead, it was dynamic. Consequently, traditional Africans were constantly adjusting and developing new architectural styles and materials to stay abreast with the changes.

This introduces into the picture another important element or dimension of pre-colonial architecture, time. Buildings have always constituted undisguised reflections of their time. Thus, the true meaning of buildings is that they are essentially expressions of their epoch (Crisman, Online). Seen from this vantage point, it becomes clear that comparisons of pre-colonial African architecture and building technology with contemporary Western or other variety are, to say the very least, unjust and smacks of insincerity. Another important dimension of architecture and building technology worth discussing is 'place.' Every building activity occurs in a specific and well-defined geographic location. This location can be appreciated at a number of different but overlapping levels, to wit, local, regional, national, or sometimes in relationship to a prominent landmark. In traditional Africa, more than elsewhere, one finds an unparalleled degree of physical consistency with the notion of 'placeness.' An example of this is the case of hill style architecture described above. Place and context often blend together to give birth to what is alluded to in architectural lingo as 'vernacular architecture.' Vernacular architecture is usually the product of specific climatic, cultural, and economic conditions of a particular place. Here, the notion of 'place' encompasses a wider geographic area – say, the regional scale or larger. In practice, an architectural style or form can be imported or borrowed from another region and once in its new locale, it undergoes a number of changes necessary to adapt it to the specific climatic, cultural, and economic conditions of the locale. The resultant architectural style or form, including the construction materials employed, is what is referred to as vernacular architecture. It is no exaggeration to state that traditional Africans were exceedingly attentive to architectural detail. In this regard, all aspects – including those that may reasonably be considered minute – were scrutinized. Only elements with significant functions, some of which may not be readily apparent to non-Africans, were included. With little, the African 'builds sensibly, functionally, and above all, beautifully' (Gardi, 1973: 248). Consider the flat roof terrace of a typical Sudanese style building for instance. This roof serves several purposes. It can be a chicken yard, a platform for drying agricultural produce, and during the long hot season typical of the Sahel, it serves as a bedroom under the stars. While such maximal utilization of space is possible with traditional housing units, it is virtually impossible with the drab tin-roofed units that have been rapidly replacing them in colonial and post-colonial Africa.

Colonialism and the demise of traditional architecture

Colonial authorities in Africa were patently interested in two intimately interrelated goals. The first was to transform the built environment in Africa into a deliberately poor imitation of its European counterpart. The second was to incorporate the continent into the global capitalist economic system. Capitalist development could not succeed whilst Africans continued to depend on building materials they could obtain free of charge from the local environment. Cognizant of this, European colonial authorities embarked on the ambitious projects that culminated in the institution of European-style physical and spatial planning laws in colonial Africa. I will concentrate here on the laws that were specifically designed to regulate the housing production process.

As far as regulating the building production process was concerned, two options were available to the colonial authorities. They could have elected to adopt performance-based standards; or they could have chosen to institute specification-based standards. For reasons that will become obvious momentarily, they elected to go with the latter, namely specification-based standards. Before delving into what this entailed, it is necessary to shed some light on the meaning and essence of performance-based building standards or codes. In practice, performance-based building codes are by design broad in their scope, and basically stipulate the desired objectives of buildings or different elements thereof, without prescribing any particular building material.

Thus, performance-based codes do not favour one particular building material over another. All what is important is that the material meets certain performance criteria. For instance, materials for the roof need to be good enough to prevent rain from penetrating the building; walling materials need to possess the ability to prevent a building's occupants from the elements, and so on. Many traditional African building materials have proven over the course of several centuries that they are capable of passing any reasonable performance test. Thus, a plethora of traditional building materials would have been deemed adequate to the task had colonial authorities adopted performance-based codes. They did not. Doing so would have gone against the grains of the objectives of the colonial enterprise.

In keeping with the broader goals of the colonial project, colonial authorities from Cairo to Cape Town and from Dakar to Mogadishu adopted specification-based building codes. These codes lacked the versatility of their rivals, performance-based standards. However, they proved indisputably effective and efficient tools of control for colonial authorities. With prescriptive building codes, these authorities were able to simply stipulate exactly what type of building material was permissible and how it must be used. In practice, this meant importing and adopting the codes that were already in force in some European municipality for use in what at the time were newly acquired colonies in Africa. There was absolutely no regard for local conditions or circumstances.

Unsurprisingly, the codes were blatantly biased in favour of building materials such as cement, steel, corrugated aluminum sheets, tiles, shingles, plywood, glass, and so on, originating in Europe. With indigenous elements taking over power subsequent to independence in most of Africa in the 1960s one would have expected some significant changes in favour of local building materials. Alas, this was not the case! Instead, the indigenous leadership in one African country after another elected to inherit the specification-oriented building codes left behind by the departing colonial authorities.

In fact, the need to become modern, nay, to appear modern, has been accentuated since the demise of colonialism. Consequently, post-colonial authorities have deemed it necessary to promulgate prescriptive building codes that are significantly more restrictive than those they inherited from their colonial predecessors. In practice, the tendency has been towards the wholesale importation and adoption of Western architectural forms and building technology in Africa. The implications of this for the development aspirations of African countries have been far-reaching. I will spend the remainder of this chapter identifying and discussing some of these implications.

Socio-economic implications of non-traditional materials

Elsewhere (see Njoh, 2000; 1999), I have proffered the following as comprising some of the leading socio-economic implications of promoting imported building materials in Africa: These implications include, but are certainly not limited to the following:

- Inequalities in the distribution of wealth;
- Unemployment;
- Housing shortage;
- Foreign-exchange difficulties;
- Technological dependence; and
- Gender-based socio-economic disparities.

Inequalities in the distribution of wealth result from policies that promote building materials originating in Europe and cognate economies while discouraging the use of traditional or locally-available varieties. The reason for this is simple. Such policies, particularly because of their penchant for capital-intensive, as opposed to labour-intensive, technology, tend to ensure that only individuals with a good level of formal training in the modern building trades are employable in the construction industry. Such individuals are invariably drawn from the small economically well-off segment of the society. This problem constitutes the source of a multiplicity of other problems, such as functional differentiation in the labour market for which African economies are well-known. Functional differentiation typically arises from the obvious and significant differences between the wage entitlements of technicians in the modern construction industry and those of their counterparts in the traditional sector.

The construction industry has an enormous potential to create jobs. Unfortunately, this strength of the industry is not taken advantage of, and in some cases, actually reversed, in Africa. The ability of the construction industry to create jobs results, at least in part, from its backward linkage mechanism, which effectively creates employment in the building material and related industries, and its forward linkage mechanism, which creates employment in sectors dealing with housing-induced consumer goods, such as furniture, household appliances and ornamental products.

Yet, the extent to which the construction industry's potential as an employment generator can be maximized is largely dependent on the type of building materials and techniques adopted. Seen from this perspective, it is clear that promoting imported materials works to the detriment of domestic substitutes and effectively weakens the construction industry's employment generation ability. According to empirical evidence I uncovered about a decade ago (Njoh, 1995), policies promoting the use of imported building materials worked to effectively discourage the creation of jobs – save a few at the distribution and assembly stages – in Cameroon's construction sector.

Another notable consequence of prescriptive standards, especially those skewed in favour of imported building materials, is housing shortage. By discouraging the use of local materials as these policies are wont to do, opportunities for developing otherwise affordable housing are eliminated. It is well established that building materials account for a substantial portion of the final cost of housing. In other words, the higher the cost of building materials, the higher the cost of construction. Labour cost tends to be negligible given its abundance in Africa. This is however, less so when building materials are of the imported genre. This is because, as stated earlier, imported building materials tend to require more formal and school-learnt skills.

Policies designed to promote imported building materials have also been known to constitute the source of foreign-exchange problems. Particularly, the excessive dependence on imported materials often constitutes a drain on foreign-exchange earnings. This is because such dependence invariably results in the loss of valuable resources, especially capital. This loss usually manifests itself in terms of the financial capital that must be paid to overseas-based building material manufacturing entities. More than occasionally, there is a need for an African country to hire expatriate technicians to execute certain specialized tasks in the construction sector. This also constitutes another source of drain on the foreign-exchange as the technicians usually require payment for their services in foreign currency. Finally, there is also the need to import the pieces of equipment necessary for assembling the modern building elements and materials. This constitutes yet another source of foreign-exchange loss.

Cultural and technological dependence have also been known to result from policies that favour foreign building materials over domestic equivalents. African policymakers cause their countries to effectively depend on foreign economies to the extent that they legislate only materials originating in these economies as acceptable for the purpose of producing something as basic as housing. Also, to the extent that housing constitutes an element of cultural expression as discussed above, African policymakers succeed in making Africans culturally dependent on Western countries by overvaluing the building materials, techniques and architecture of these countries while devaluing African varieties.

The cultural implications of supplanting African building and architectural practices with Western varieties cannot be overstated. Building technology, like other technologies, is laden with a cultural baggage. Accepting any given technology invariably implies accepting the attendant cultural baggage. Witness for instance, one consequence of supplanting Africa's construction processes, in which women were very active, with European varieties, wherein women were absent. The consequence is that women are absent or terribly underrepresented in the modern construction sector in contemporary African countries. The exclusion of women from the modern construction sector has served to widen the gap between the earning potential of men and women in Africa. This is because a significant proportion of the better-paying jobs in African economies can be found in the construction sector. This constitutes the basis of the charge that adopting foreign building materials in the construction industry has resulted in gender-based socio-economic disparities in African countries.

Chapter 11

Development Implications of Tradition

Introduction

The discourse on tradition and development has gone beyond wrestling with the question of 'whether tradition matters' to 'how tradition matters' in the development process. These questions were absent from the same discourse when Western countries were at about the same stage of development as contemporary African countries. There has never been any question as to whether or how Western arts, music, dance, and languages affect development. In fact, the theory of a positive association between tradition and such desirable outcomes as improvement in skills and talents, knowledge, community cohesion and socio-psychological development is taken as a given in the West. It is therefore hardly any wonder that much time, energy and financial resources are dedicated to the development, maintenance and/or preservation of relics and vestiges of Western culture and civilization.

Tradition, culture and development have always been considered inseparable in Western thought. This is especially true when the term development is taken to be synonymous with 'modernization.' Seen from this perspective, as members of the international development community were wont to erroneously do in the 1950s and 1960s, development is taken to constitute part of Western culture or tradition. This view led to the wholesale recommendation for countries in underdeveloped regions, particularly Africa, to abandon their customs and traditional practices in favour of Western substitutes. This is because African customs and traditional practices were considered incompatible at best, and antithetical at worst, to development. Thus, for members of the international development community of that era, the fate of African countries appeared sealed by the nature of their culture and tradition. This view, as Amartya Sen (2004) would say, was not only a 'heroic oversimplification' but it also entailed some assignment of hopelessness to African countries simply because they did not have the right kind of culture or tradition. Sen is even more blunt when he lambastes such a view for being not only 'politically and ethically repulsive' but also tantamount to 'epistemic nonsense' (Sen, 2004: 38).

Even today, few Westerners are objective enough to appreciate the folly and absurdity of this recommendation. In fact, equally few Westerners find anything rational about African cultures and traditional practices. If you are a Westerner and have taken the time to read this book, it is highly probable that you are one of the few. For most, these cultures and traditional practices are 'primitive' and symptomatic of 'backwardness.' For those bent on universalizing the capitalist ideology, Africans cannot develop unless and until they have abandoned their so-called 'primitive' traditional practices and institutions. This is because such practices lack the social change mechanisms necessary for attaining contemporary development objectives. From this vantage point, the cultural and traditional foundations of social relationships in Africa are irrelevant to modern socio-economic development.

My aim in this concluding chapter is to challenge this viewpoint and argue that African cultural and traditional practices constitute a sound – perhaps the only sound – foundation upon which the structure of any meaningful socio-economic development strategy for Africa must repose.

Tradition, culture and development

Helen Gould and her colleagues of the the British-based NGO, Creative Exchange has proposed a useful three-tier scheme for discussing the notion of culture in the context of development. These include (Creative Exchange online), culture as content, culture as context and culture as method. Culture as context has to do with the cultural environment of development projects. For instance, a development project may constitute an affront or challenge to local cultural norms, traditional practices and customs. Here, it is important to note that some of what has come to be erroneously seen as African custom, such as so-called traditional gender roles, actually have their roots in the European colonial era. Culture as content has to do with drawing on cultural and traditional practices to inform contemporary development endeavours. For instance, contemporary development planners can draw lessons or items from traditional rituals such as initiations into adulthood, to facilitate accomplishments of development goals. Finally, culture as method has to do with the use of cultural expressions, including, but not limited to, songs, dance, poetry, idioms, and proverbs to enhance development efforts. In this case, culture may be employed in one of two ways. First, it may be used as a tool, which can generally be message or content-oriented. Second, it may be employed as a process, particularly entailing not only the devolution of power from the center to the provinces, but also, and more importantly, reinforcing local people's control over the development process. To be sure, not all cultural elements or traditional practices have any utility for contemporary development initiatives on the continent. In fact, traditional practices such as female circumcision, and activities that go against the grains of development such as the use of children in armed conflict, that claim as their foundation the African tradition of maintaining 'age-sets,' have no place in modern Africa. When traditional practices infringe on basic human rights, the traditional practices must yield. This, essentially, is the case with female circumcision and the recruitment of children – 'child soldiers' – into the military.

Therefore, the need to identify African traditional practices that can facilitate the attainment of critical development objectives in contemporary Africa cannot be overstated. In the remainder of this chapter, I seek to show how African traditional practices can be incorporated into strategies to:

- Promote African culture as a means of facilitating attainment of other development goals;
- Promote development of an appropriate development ideology;
- Improve ecological management and agricultural productivity; and
- Promote regional integration.

Culture and the attainment of other development goals

African culture is often treated as a static phenomenon. Yet, African culture, like Western or any other culture for that matter, has never been static. Perhaps more importantly, African culture, like other cultures around the world, contains attributes that have been borrowed from others. To be sure, no society can claim that its culture is pure and unadulterated by outside influences (Gyekye, 1997). However, it is foolhardy and in fact, counterproductive for a society to abandon its own cultural practices in favour of alien varieties as a way of changing for the sake of change alone. The borrowing of an alien cultural practice is only logical when it has a comparative advantage over the indigenous variety it seeks to replace. The notorious and persistent efforts to supplant African culture with Western varieties have never relied on such logic. Rather, the aim has consistently been to promote economic development in the metropolitan countries as well as assert the supremacy of Western culture and civilization. Yet, African countries stand to benefit tremendously in social, economic, psychological and cultural terms by promoting appropriate and relevant African traditional practices in all domains. Thus, there is a need to protect, promote and preserve such practices as an objective of development on the continent. To appreciate this line of thought, it is necessary to understand five important contributions of culture to development as elaborated by Amartya Sen. Sen (2004: 39-42) contends that culture may be:

- A constitutive part of development;
- A set of economically rewarding activities;
- Influential on economic behaviour;
- A facilitating force to political participation; and
- A facilitator of social solidarity and association.

To Sen's list, I add the ability of culture to facilitate efforts to,

- Improve housing conditions.

I cull substantive examples from my discussion of traditional African architecture and building techniques in Chapter Ten to ground the otherwise abstract argument in this section.

Culture as a constitutive part of development

To appreciate culture as a constitutive part of development, one must begin by addressing the question: what is the purpose of development? As I mentioned in Chapter One, culture is at once a means and end of development. As a means, culture can facilitate the attainment of other development goals such as economic, social and political progress. As an end, cultural development reposes on achievements in other domains, particularly a viable economy. For example, the preservation of architectural artifacts from the past, even if they possess no immediately apparent economic value, requires economic resources. As a means to an end, development planners and policymakers may advocate traditional practices, particularly because they are more cost-effective in comparison to imported substitutes. Consider for instance, the case of traditional building materials and construction processes. As discussed in Chapter Ten, these materials and techniques are not only far less expensive, they are also more environmentally friendly, than the imported brands that Western change agents have been promoting on the continent since the colonial era. Therefore, *ceteris paribus*, efforts aimed at protecting, preserving and promoting African traditional practices and culture invariably result in improving living standards on the continent. The case of indigenous architecture and building techniques in efforts to address the problem of qualitative and quantitative deficiencies in the continent's housing stock is illustrative. As argued in Chapter Ten, reducing the cost of building materials by employing local varieties goes a good way in reducing housing cost. This is especially because building materials account for a disproportionately large proportion of the cost of new construction.

Culture as a set of economic remunerative activities

Contrary to popular opinion, the preservation of cultural artifacts and sites can also pay economic dividends. The hundreds of thousands of visits that are paid to sites of cultural significance in Egypt, Israel, Ethiopia and other countries with a long and rich history are illustrative. Many a visitor to Egypt wants to have the opportunity to examine the pyramids close-up. Visitors to Ethiopia want to see and admire the ingenuity that went into carving housing units literally into rocks. Visitors to Israel are often desirous of seeing first-hand the birthplace of Jesus Christ. Muslims from all the four corners of the world make annual pilgrimages to Meca to see Islamic sites of great antiquity. No wonder therefore that tourism is often linked to the cultural environment, the ethical and other questions associated with commercializing historic and cognate sites notwithstanding. The decision to preserve African indigenous cultural and traditional artifacts must therefore be seen as a decision having implications not only for cultural but also economic development. Take my call to protect, preserve and promote traditional and cultural elements of African architecture for example. The economic value of these elements can be appreciated at three different but overlapping levels as follows. First, it can be seen in terms of the positive economic externalities associated with tourism. Second, the economic importance of African traditional architecture manifests itself in terms of the savings accruing from inexpensive local building materials as opposed to the expensive imported varieties. Finally, there are the earnings deriving from increased employment in the local construction sector. Here, it is important to note that the jobs will be for men and women, as well as individuals from all socio-economic classes. This is because the skills necessary for functioning in the informal or traditional construction sector are, in contrast to those required in the modern sector, not school-learnt.

Culture and economic behaviour

Human behaviour is a function of the human environment. A close examination of African traditional practices and culture reveals a sophisticated understanding of this relationship. For instance, the design of space within the framework of traditional architecture seeks to encourage positive and productive behaviour. Accordingly, the typical traditional African compound is designed in a manner that conveniently accommodates the extended family, ensuring that every member and activity is guaranteed room. There is space for storing grains, and space for the preparation of food and space in which women are guaranteed some privacy, and from where they can see but not be seen. In regions such as the Sahel, roofs double as sleeping space during periods of the year when interior space is unbearably hot at night. In this case, the aerated roof terrace affords people the opportunity to have a restful night and to be rested enough to attend to socio-economic demands of the day.

Culture and political participation

Culture affects in no small way the extent to which people can, and do participate, in the political process. In this regard, one of the most influential aspects of culture has to do with the design and structure of the built environment. European colonial authorities did not facilitate participation in the political process by the spatial structures they created in Africa. In colonial Africa, power, as manifested through government buildings and other infrastructure was located on hilltops, as far away from the people as possible. The hilltop location ensured, or at least gave the impression, that the 'natives,' who were always located at the foot of the hill, were literally under the constant gaze of the colonial authorities. Thus, the choice of such sites had multiple objectives, with the need to buttress and assert the colonial government's control over the 'natives' standing out as primordial. The colonial governor's mansion was typically enclosed within an intimidating fence complete with armed guards. The colonial authorities and the indigenous leaders, who inherited the helm of government from them, maintained and reinforced this spatial structure thereby, effectively contributing to discouraging political participation.

In contrast, pre-colonial African architecture and spatial structures sought to promote political participation by locating the chief's, king's or other traditional leader's home, which was fondly referred to in most African societies as 'the people's home,' in the heart of the village, town or hamlet. For any given human grouping in traditional African society, the leader's home constituted the geometric center of that grouping. In traditional Yoruba society, the *Afin* or Oba's or King's palace was so centrally located that all roads were said to lead there. In traditional Yorubaland, the main market of any given village was always located in front of the palace of the chief or king of that village. Each palace contained several courtyards, with the largest one serving as the venue of meetings or gatherings of the townspeople. The palaces of kings, chiefs and other traditional African leaders served as the refuge for strangers, visitors and others who were rendered temporary homeless. Thus, as mentioned above, the palaces were fittingly referred to as the 'people's home.' In the non-centralized or stateless societies of Africa, decisions on matters of societal importance were taken in public squares – usually under large centrally-located trees.

Culture, social solidarity and association

As Sen (2004) contends, and I concur, the operation of social solidarity and mutual support is a function of culture. The success of social relationships depend largely on societal norms governing what people can do for others or expect others to do for them. For instance, in the construction sector in precolonial Africa, people were expected to cooperate, through rotating task execution groups, and take turns building housing units for one another. Faced with problems of resource constraints, these types of arrangements continue, in contravention of government rules, in urban centers and especially at the periphery of these centers as well as rural areas throughout Africa.

Culture and efforts to improve contemporary housing conditions

African governments dating back to the colonial era have actively sought to discourage the use of traditional building materials and informal processes in the construction industry. In effect, these governments have been bent on supplanting informal processes with more formal varieties. In practice, this has meant amongst other things, requiring only licensed, formally trained builders to undertake building projects. Concomitant with this, is the requirement that no building can be occupied unless it has been completed and inspected by building professionals. Furthermore, and most important of all, the governments have systematically adopted specification-oriented, as opposed to performance-oriented, standards.

As discussed in Chapter Ten, these policies have had serious negative impacts on the housing sector in African countries. On account of the growing deficiency in the housing stock of these countries, there is no denying that an urgent need to seriously revamp these policies exists. The logical starting point in this regard is to tap lessons from the past, when individuals produced their own housing using locally available materials. This is not to advocate the complete abandonment of standards. Rather, it is to advocate the adoption of performance-based standards.

Some, such as J. Kanyemba of the Rural Industries Innovation Centre in Kanye, Botswana, have hinted at some of the difficulties inherent in this proposal (see Kanyemba, Online). The primary difficulty, it is argued, has to do with the structural and other tests that would have to be conducted on the various traditional building materials in order to ensure their soundness. This difficulty appears overblown given the fact that all the known traditional building materials in Africa have been in use for centuries. Thus, the materials in question have been subjected to centuries of tests and re-tests. It is therefore safe to conclude that the strengths and weaknesses of the materials are by now common knowledge. African governments and interested parties in the international development community will do well to concentrate on taking advantage of the strengths, while devising strategies to eliminate the deficient features, of the materials.

Another difficulty has to do with how to word performance-based codes. Again, this difficulty is exaggerated. Let me pause here to distinguish between performance-based and prescriptive building codes. While prescriptive codes state in very definite terms the specific building material that must be used, performance-based codes stipulate only the objectives that must be attained while allowing discretion on the selection of the specific type of building material to the builder. The relative difficulty of crafting performance codes has to do with articulating the standards that buildings or elements thereof are required to meet.

On closer examination however, this turns out to be an issue that ancient Africans had wrestled with and resolved centuries ago. For instance, building units in areas, such as the Sahel belt of West Africa, where temperatures were known to vary significantly between the day and night, used materials and building designs that protected the building interiors from virulent winds, and cold temperatures during the night and the intensive heat of the day.

Tradition, culture and development ideology

As noted throughout the previous chapters, Christian missionaries and colonial authorities were at the forefront of efforts to supplant African culture and traditional practices with Western varieties during the colonial era. The efforts of Christians in this connection did not cease with the demise of colonialism. To be sure, these efforts are ongoing. Today, Africa tops the list of regions in which Christianity is experiencing its highest growth in the world. It is in order to note at this juncture that apart from Christianity, other non-indigenous religions, particularly Islam, have made significant in-roads in Africa over the years. However, the record of other non-indigenous religions in destroying African culture and traditional practices pale in comparison to that of Christianity. It is particularly for this reason that I elect to focus specifically on Christianity and its impact on African tradition and development aspirations.

A clash of ideologies

Christianity has always avowedly supported and promoted the twin Western ideologies of capitalism and individualism. Max Weber is among the most notable thinkers in modern history to draw attention to this phenomenon. Weber (1905) observed that Christianity, especially Protestantism, emphasized thrift, discipline, hardwork and individualism – all hallmarks of capitalism. This is what came to be widely known as the Protestant ethic. Weber associated the tendency for capitalist orientation with Protestants based on his observation of religion and society in Europe in the early–1900s. Weber (1905) noted for instance that, business leaders, members of the professional and technical class and other economically well-to-do individuals in Germany at the time were overwhelmingly Protestants. The dominance of Protestants within the higher socio-economic echelons of society was also true of other European societies. Weber appeared uncertain about why this was case, and pondered whether religious affiliation was a cause or the result of material well-being. Weber however, had no doubt that modern capitalism had flourished largely in those areas of Europe where Protestantism had taken root early in the Protestant Reformation.

In Africa, especially during the colonial and immediate-post-colonial époques, Christians of all stripes disproportionately enjoyed the fruits of modern capitalist development. The link between Christianity and success in the imperial capitalist economy is easy to understand. As mentioned in Chapter Three, access to modern education was one of the most notable ways in which Christian converts in colonial Africa were rewarded. Modern education in turn permitted access to jobs in the formal sector, which invariably guaranteed access to material and other resources. However, it must be noted that women were conspicuously absent from the formal sector of the economy in colonial and immediate-post-colonial Africa. This was certainly no accident. Rather, it was a function of the fact that Christian and colonial authorities deliberately assigned no significant roles to women.

Western change agents focused on material gains as a measure of development and prosperity. In concert with this orientation, development planning efforts in Africa were typically evaluated by concentrating on indicators of economic progress such as gross national product (GNP) and gross domestic product (GDP). This exclusive focus on economic development invariably ignored other aspects of development. For instance, no effort was made to determine the extent to which women and other disadvantaged groups were systematically excluded from the public sphere. Yet, as I have mentioned throughout this book, women were active both in the public and domestic spheres in pre-colonial Africa. Western analysts have been wont to ignore or distort this fact.

However, contemporary African feminists such as Ifi Amadiume (e.g., 1987) and Niara Sudarkasa (e.g., 1986) have embarked on efforts to set the record straight by challenging received orthodoxies that have interpreted this and other aspects of African culture in the image of Westerners. If nothing else, the exemplary works of these scholars have succeeded in painting a picture of traditional or pre-colonial

Africa that approximates reality in ways Western social anthropologists would never have imagined. As a result, it is becoming increasingly clear that life for women in pre-colonial Africa is far-removed from the chauvinist stereotypes created by colonial anthropology. Apart from the fact that sex and gender did not coincide in pre-colonial African societies, these societies had structures that permitted women to ascend to positions of power and authority. More importantly, societal roles were neither rigidly masculinized nor feminized. To be sure, many African societies were patriarchal. Similarly, a good number of these societies were matriarchal – a fact that systematically and consistently eluded colonial anthropology. Colonialism, with its patriarchal proclivities did well to eliminate the matriarchal societies and universalize patriarchy. These developments have been to the detriment of the socio-economic development of women throughout the continent. Western feminists and other Western change agents concerned with improving the lot of African women appear oblivious to this fact. Rather, these agents persist in their incrimination of traditional African culture as the source of the African woman's social, economic and political problems.

Christianity's emphasis on capitalism and individualism, and its tendency to relegate women to subordinate roles, conflict sharply with African traditional religious beliefs and practices. Women have always played important roles in traditional African religion, as priestesses, diviners, rainmakers, and healers amongst others. African traditional religious doctrine also discourages individualism and emphasizes the spirit of collectivism. It places the interest and welfare of the group (extended family, lineage, clan, village, hamlet, and so on) over individual interest. The sense of community is a well-known feature of African life (Gyekye, 1997). The pronoun 'I' is seldom used in public assemblies throughout traditional Africa. 'By emphasizing communal values, collective goods, and shared ends, a communitarian social arrangement necessarily conceives of the person as wholly constituted by social relationships' (Gyekye, 1997: 37). Within the framework of African religious thought and political philosophy, certain goods are considered unattainable unless treated in the context of the shared life of society as a whole. Critics are likely to contend that the communitarian nature of African culture means individual rights and liberties are neglected in favour of communal rights. However, it is important to note that as defined by Pope John Paul II in his 1987 encyclical on social concern, *Sollicitudo rei socialis*, the proper good of the human person *(bonum humanum)* is closely related to the common good of all *(bonum commune)*. Thus, the proper good of individuals does not have to be sacrificed to attain the common good of society, as zero-sum theorists would have us believe.

The supplanting of African indigenous religion with Western varieties, particularly Christianity also has implications for economic development in Africa. Christian doctrine designates one day per week on which no (manual) work must be carried out, and several days per year that must be observed as Christian holidays (see Chapter Three). This feature of Christianity has far-reaching negative implications for development efforts in Africa. This is because most work, in Africa in particular, and regions at a similar level of development in general, is of the manual variety. The problem is magnified when we take into account the fact that most farm or agricultural work is seasonal. This means that certain activities must be performed during certain periods of the year and every day during some specific seasons (e.g., planting and harvesting seasons) matter. This is because tasks that must be accomplished during a certain period, say planting, cannot be deferred to another period. Crops must be planted during a specific period, otherwise they would not germinate; also, they must be harvested during a definite period otherwise they would rot in the farms.

Culture and ecological management and agricultural productivity

Several factors account for the dismal record of development planning in Africa. However, few of these factors rival the tendency on the part of development planners to ignore the wealth of knowledge that Africans possess about their own problems. This wealth of knowledge, which constitutes a part of pre-colonial African tradition, falls under what is known in the relevant literature as indigenous knowledge. As the discussion in this book demonstrates, pre-colonial Africans were vastly knowledgeable in many areas. In this section, I discuss African indigenous knowledge in a number of areas deemed critical to ongoing efforts to promote sustainable forestry and natural resource management on the continent in particular and the world in general. Particularly, I discuss indigenous knowledge (IK) in the domains of natural environment and agriculture, with a view to underscoring the need for development planners to pay more attention to local people's knowledge of possible remedies to some of the continent's most nagging problems. The conventional practice is for development planners to harbour an entrenched superiority complex that prevents them from learning from those for whom they profess to plan. Thus, development planners have typically proceeded as if they possessed infinite knowledge of the problems they are tasked with resolving while those who are directly impacted by the problem know nothing.

Sustainable forestry and natural resource management

Forestry and natural resource management is one of the many areas in which development planning efforts have been greeted with minimal success in Africa. Again, these disappointing results have been due to, amongst other things, the fact that planners and policymakers have consistently ignored centuries of indigenous knowledge. Yet, as Warren (1992) has noted, the indigenous knowledge and organizations of Africans can serve as the foundation for cost-effective and sustainable participatory approaches to forestry and resource management. Some of the knowledge I have in mind falls under what has come to be known as traditional ecological knowledge (TEK). Traditional ecological knowledge can be defined as knowledge and beliefs about the natural environment that has been handed down through generations by cultural transmissions (Mathooko, 2000). Africans have lived for centuries in harmony with their forests and concomitant natural resources. For the African, the forest, like land, was never recognized for its economic worth. Rather, the forest was recognized for its spiritual, religious, medicinal, and cultural significance. It was also recognized as an important source of food. In this latter regard, Africans had a complex set of laws governing the type and quantity of resources that could be obtained by individuals and/or whole communities. In addition, Africans had developed an integrated system of knowledge, practice and beliefs to accompany these laws. Most importantly, Africans saw themselves as custodians as opposed to owners of the forests. As custodians, the living recognized it as their duty to ensure that they bequeath to unborn generations a healthy inventory of forests and associated natural resources. To ensure that this came to pass required a high degree of ingenuity and wealth of traditional ecological knowledge (TEK).

Traditional ecological knowledge promises to be of enormous benefit to ongoing sustainable forestry and natural resource management efforts in Africa. Mathooko (2000) has identified and discussed some examples of TEK in the area of aquatic resource management in Africa. Over the years, local fishermen in Africa have developed different coping strategies. Those in areas with an undersupply of fish employ a strategy akin to that used by farmers who occasionally fallow land in order to permit the land to rejuvenate itself. In the case of fishing, the fishermen occasionally restrict fishing to certain specific areas while allowing others to regenerate resources. Also, traditional fishermen have over the years, learnt to compensate for poor harvests seasons by increasing fishing activities in other areas. As Mathooko (2000) observed, such flexibility is a function of TEK about appropriate substitutes for scarce resources. Such knowledge, including knowledge about the spatial distribution of different species and quantity of fish has been passed down from one generation, and/or from one fisherman, to another. Africans had developed relatively accurate 'mental maps' of aquatic resources long before the European conquest and a lot longer before the advent of sonar technology. Traditional African fishermen possess the ability to detect the location of different aquatic life, including rare and endangered species by simply tasting mouthfuls of water from the general area or by submerging their heads to listen for specific types of noises in the water. Even with the availability of sonar and other sophisticated technology, local knowledge of the locations of threatened aquatic life, fish migration routes and aggregation sites remains more reliable.

Other areas of aquatic resource management in which indigenous ecological knowledge has proven invaluable are as follows (Mathooko, 2000): provision of new biological and ecological insights into aquatic ecosystems (e.g., life cycle tropical reef fish); management of wetlands; protected areas and conservation education; evaluation of aquatic and natural resource production systems; and environmental assessment.

Traditional methods for boosting agricultural productivity

Food insufficiency has been a recurring problem in Africa since the nightmarish famine of Ethiopia in the late 1970s and early 1980s. One reason why efforts to combat the problem have failed to yield significant positive results is linked to the tendency on the part of development planners and policymakers to ignore indigenous African knowledge on agriculture. Yet, Africans have over the period of several centuries, developed many innovative, effective and efficient techniques for boosting agricultural productivity in even the most precarious environments. I describe three such techniques, including the Zai technique for recovering crusted land; the use of vegetables to improve soil quality; and traditional techniques for improving animal breeding.

The Zai technique

Crop production in many parts of Africa, particularly the Sahel region, has not kept abreast with population growth since the 1980s. This has caused, among other things, increases in levels of poverty, hunger and starvation in several areas. Efforts on the part of national and international authorities employing so-called modern farming techniques to reverse the dreadful situation during the last two or so decades have registered mixed results. Recently, the attention of concerned authorities has been drawn to Burkina Faso, where local farmers are employing, with remarkable success, an ancient indigenous technique to restore, maintain and improve soil fertility known as the Zai technique. In applying the Zai technique to recover crusted land, farmers dig a planting pit measuring between 20 and 40 centimeters in diameter and 10 to 20 centimeters in depth, depending on the soil type. The pits, which are partially filled with organic matter (about 0.6 kg/pit), are dug during the dry season (November – May). Subsequent to the first rainfall, the organic matter in each pit is covered with a thin layer of soil. Then, seeds are planted in the middle of each pit. Finally, the excavated earth is ridged around the mouth of the pit as a means of enhancing the pit's water retention capacity. The number of pits per hectare, usually vary from 12,000 to 25,000. The reasons for the technique's success stem from the fact that it ensures, *inter alia*, that,

- As much water as possible is captured from the rain and surface run-off;
- Seeds and organic matter are protected from being washed away;
- Yields are significantly increased;
- Water and nutrients are available at the beginning of the next rainy season; and
- Biological activities in the soil are reactivated thereby improving the soil structure;

In sum, the Zai technique fulfills three critical functions in agriculture in semi-desert areas, namely soil conservation, water conservation, and erosion control for otherwise infertile soils.

Using traditional vegetables to improve soil quality

Traditional vegetables are valuable not only as food but also as a source of medicine in Africa. This is about as much as non-Africans have known for decades. However, in the recent past, members of the international development community have begun embracing a new reality, namely the fact that traditional Africans use indigenous vegetables to boost agricultural productivity. In this regard, traditional Africans have used indigenous vegetables to compensate for the lack of modern agricultural inputs such as fertilizers and pesticides. Instances of the use of indigenous vegetables for this purpose were registered by researchers working with farmers in a parish located about 50 km west of Kampala, Uganda as recently as June 2002. The farmers indicated that they used manure or rested the soil, as a strategy to improve soil quality and restore nutrients. However, this latter method appears not to be very useful because of the intensive farming systems employed by farmers in the region. This is where the use of indigenous vegetables came in. In the 1970s, a number of local traditional farmers discovered that the cultivation of some indigenous vegetables had the effect of vicariously improving soil quality, thereby nullifying the need for specially prepared manure or resting the soil. Since making this discovery, local traditional farmers have adopted the habit of rotating various exotic vegetable crops with traditional varieties as a means of improving the yield of the former. The gains in yields realized in the process are usually superior to those associated with the use of conventional fertilizers and manure. For instance, the rotation of green beans *(Phaseolus vulgarus)*, Ebugga *(Amaranthus dubius),* and tomatoes *(Lycopersicum lycopersicon)* in that order, has been to result in Ebugga adding beneficial properties to the soil, thereby causing improved yields by the other two crops.

Indigenous animal breeding techniques

Animals have always been vital to Africans. Animals serve as food and are also used for religious and ritualistic purposes. Over the course of several centuries, Africans, especially members of the African pastoralist community, have developed different methods for not only conserving breeds and animal diversity but also of influencing the genetic composition of their herds. The reasons for this are plentiful but include the following main ones (Practical Action, Online):

- Cultural norms regarding the use of animals;
- Local preferences for certain attributes, such as colour, size, and behaviour;
- Selection practices for certain qualities (culling, offspring testing);
- Pedigree-keeping; and
- Social restriction on the sale and/or exchange of animals.

Resulting from centuries of experience have been several breeds of animals in Africa. By some account, there are as many as 7000 breeds, 'each one specialised for a particular purpose. Each animal and each breed carries with it a particular set of genes – a code that determines everything from the animal's colour to its milk-yielding ability; from disease resistance, to the likely number of offspring' (Practical Action, Online). Contemporary development planners cannot afford to commit the grave mistakes of their predecessors by ignoring valuable indigenous knowledge that has been accumulated through centuries of experience on the continent. Indigenous knowledge about animal breeding can serve as invaluable source of information about hitherto scientifically undocumented breeds and their adaptive traits. For instance, ancient Maasai had developed a special breed of sheep, commonly known as the Maasai sheep. This fat-tailed sheep possesses unique characteristics that render it resistant to internal parasites. Knowledge of these unique characteristics promises to be of enormous value not only to animal breeding efforts in other parts of Africa but the world as a whole.

Tradition, culture and regional integration

One reason why African countries south of the Sahara have lagged behind the rest of the world has to do with their diminutive sizes. The colonial powers, authors of the mosaic pieces of territories found throughout the region paid no attention to politico-economic development as a national objective. Rather, they were concerned with minimizing the cost of resource exploitation and maximizing profits for the metropolitan countries. In terms of geography and population size, some African countries, such as Equatorial Guinea, Gambia, Sao Tome and Principe, Cape Verde, Lesotho, and Swaziland, to name just a few, are smaller than many major American cities. The problem of socio-economic underdevelopment is certainly not confined to the smaller sub-Saharan African countries. Rather, all countries in the region, with the possible exception of South Africa, face nagging political and/or economic problems. One acknowledged potentially viable solution to this nightmarish situation is regional economic and political integration. In this section, I contend that African tradition and culture contain elements that can be culled to enhance ongoing efforts to reverse this undesirable situation.

Culture and macro-political and economic development

The notion of integration as a strategy for strengthening Africa's politico-economic potential is certainly not novel. In fact, proposals for integrating Africa date back to the colonial era when the eccentric pan-Africanist, Cecil Rhodes, expressed his cynical desire to unify Africa from Cape Town to Cairo under the Union Jack. The earliest and most notable effort towards integration in post-colonial Africa dates back to 1963, when the Organization of African Unity (OAU) was established. The creation of the OAU effectively marked the beginning of an idealistic pan-Africanist movement. A noteworthy upshot of this movement is the birth of the African Union (AU) on 9 September 1999. On this date, the Heads of State and Government of the OAU issued a Declaration, the Sirte Declaration, which called for the creation of an African Union. According to this Declaration, the AU is charged, *inter alia*, with the task of facilitating the integration process in Africa as a means of: 1) enabling the continent to enhance its global economic position; and 2) addressing the continent's social, economic and political problems, which have been exacerbated by the forces of globalisation.

To be sure, no one ever said the African pan-Africanist movement has ever been short on grand pronouncements and ambitious plans. The problem however, is that these plans and pronouncements have never been translated into reality. Witness for instance, the following major plans and programmes that were initiated during the heydays of the OAU, the forerunner to the AU (Africa-Union.org):

- The Lagos Plan of Action (LPA) and the Final Act of Lagos (1980), which incorporated programmes and strategies designed to accelerate self-reliant development and co-operation among African countries.
- The African Charter on Human and People's Rights, which was adopted in Nairobi, Kenya in 1981, resulting in the creation of the African Human Rights Commission located in Banjul, The Gambia.
- The African Priority Programme for Economic recovery (APPER) of 1985, an emergency programme which had as its main purpose, addressing Africa's development crisis of the 1980s.
- The OAU Declaration of 1990, which acknowledged the need for Africa to be in charge of its own destiny, particularly in terms of devising strategies to address the challenges of peace, democracy and security on the continent.
- The Charter on Popular Participation of 1990, which was designed to accentuate the need to involve the citizenry in the decision-making process.
- The Treaty establishing the African Economic Community (AEC), also known as the Abuja Treaty of 1991, whose avowed aim was eventually to create an African Common Market, with Regional Economic Communities (RECs) as building blocks.
- The Mechanism for Conflict Prevention, Management and Resolution of 1993, which was basically an articulation of the determination of the African leadership not only to resolve conflicts, but also to promote peace, security and stability in Africa.
- The Cairo Agenda for Action of 1995, which sought to re-launch Africa's political and social development programme.
- African Common Position on Africa's External Debt Crisis of 1997, which had as its main objective, crafting a strategy for resolving the continent's external debt problems.
- The New Partnership for Africa's Development (NEPAD) which was adopted at AU Summit in Lusaka (2001)

Apart from the Organization of African Unity (OAU) and later, the African Union (AU), a number of other entities have also been involved in efforts to integrate the African continent, or regions therein. Prominent in this regard have been the effort of regional leaders, such as those within the West African region, who established the Economic Community of West African States (ECOWAS) in 1975, those of the southern African region, who created the Southern African Development

Community (SADC) in 1979, the Economic Community of Central African States (popularly known under its French acronym, CEEAC), which was constituted in 1981, the Sahelo-Saharan States (CENSAD), established in 1998, and the East African Community (EAC), whose creation dates back to the colonial era. These entities can best be described as regional political pacts. The strongest and most ambitious of these is the East African Community, which has political integration as its avowed objective.

Unlike the others, the EAC's institutions include a high court and parliament. Apart from these political institutions, there are a number of regional groupings, which overlap political pacts. A number of these serve financial or economic functions. Included in this category are the CEMAC and the West African Economic and Monetary Union (UEMOA). These institutions combine to constitute the Communauté Financière Africaine (CFA). The CFA has been around since 1948 when, as a currency for the erstwhile French colonies, it was pegged against the French Franc – an arrangement that lasted until the advent of the Euro. Today, the currency is pegged against the Euro, which has infuriated some members of the European Union.

The record of regional integration

As I said earlier, and as the foregoing catalogue of integration initiatives demonstrate, there has been a plethora of efforts aimed at integrating African countries. However, to initiate efforts designed to integrate countries is one thing, and for these efforts to succeed is a completely different issue. The results of integration efforts in Africa, especially during the post-colonial era have left much to be desired. In fact, by most accounts, the efforts have failed woefully (see for example, Adotevi, 1997). Why has this been the case? There are several possible reasons. However, the most persuasive explanation for the inability of integration to register desirable results relates to tendency on the part of these efforts to ignore Africa's pre-colonial history while overemphasizing its colonial foundations. In his contribution to the edited text by R. Laverge (1997) Stanislas Adotevi refers to post-colonial integration efforts as 'false solutions' to the continent's problems. He contends that the failures of these efforts have been caused by three major phenomena. First, the efforts have suffered from a remarkable degree of 'reductionism and purity.' This problem is manifested by the fact that regional blocs have typically been established through sheer political will and along colonial linguistic lines or geographic boundaries with the aim of attaining exclusively economic goals. Second, there has been the tendency to venerate the nation-state and its concomitant institutions and structures. Yet, the notion of the nation-state in contemporary Africa as created by colonial powers during the Berlin Conference of 1884/5 is largely an arbitrary phenomenon. Finally, there has been the tendency to impose regional integration from above and sometimes, from outside of the continent. By so doing, popular practice and long-standing relations among peoples dating back to the pre-colonial era are ignored. Consequently, the structures of regional blocs have been built on unstable foundations. This, and the natural limits of a strictly economic approach to integration, according to Adoveti (1997) are two of the leading factors accounting for the structures' failure.

It is true, as noted in Chapter Seven that the notion of statehood pre-dated the arrival of Europeans in Africa. However, it is important to note that pre-colonial African states in no way resemble their contemporary counterparts, which as noted above, were arbitrarily created by European colonial powers, who employed solely geographic features as international boundaries. Rather, pre-colonial African states comprised multiethnic federations. Adotevi (1997) sheds some light on the role of the central authorities in these indigenous African states. These authorities had as their main function, maintaining law and order, defense, and in some cases, the collection of taxes. They eschewed any attempt at meddling with social relationships, thereby making it possible for different sub-groups to preserve their own sub-culture. Cultural integration was never an objective. Resulting from this was a rich patchwork of cultural groups with identical goals.

European colonial powers made no effort to preserve the pre-colonial federations, which transcended the new international boundaries. As a result, many pre-colonial federations or kingdoms were forcefully broken up as the colonial boundaries divided them amongst different colonial powers. This is one reason why some analysts such as Adotevi (1997) have referred to the contemporary nation-state in Africa as an artificial entity. What makes it artificial is not only the fact that they are demarcated solely by geographic features such as rivers, mountains ranges, meridians and parallels, but also the fact that they were created by non-indigenous authorities. It is especially for this reason that Africans, dating back to the colonial era, have always seen the so-called modern state as an alien entity. Thus, unscrupulous state employees who embezzle government funds consider it simply their own way of getting even with this alien entity. The average African in colonial and post-colonial Africa has always seen the state as coercive, oppressive, domineering and tyrannical. In contrast, pre-colonial Africans saw the state as an entity that guaranteed their security and ensured the preservation and equitable distribution of scarce resources.

Culture as a basis for regional integration

One exception to the unsuccessful record of regional integration in Africa is that of the Franc CFA Zone (Adoveti, 1997). Although the Franc Zone, comprising two branches, the West African Monetary Union and the Bank of Central African States, has been criticized for overvaluing the Franc CFA to the detriment of the member states, it has largely been a success. To be sure, the franc zone has been the only consistently functional bloc, something that cannot be said of the other blocs in the region. One reason for the success of the franc zone, as Adoveti, observed is its grounding in historical reality – the member countries (with the exception of Equatorial Guinea and Sao Tome and Principe) are former French colonies. As a result, the member countries have a common business language, similar traditions of business administration and fiscal management as well as a corps of elites who share a common educational background. Lest we forget, the shared colonial history of the member states of the Franc Zone, which arguably accounts for the zone's success, is of relatively recent vintage. This leads Adotevi (1997) to contend that if such a fairly short history of shared experience can account for as much success, then the centuries of shared historical experiences of indigenous African societies must be considered a sound foundation on which to build integration units for all of Africa in the 21st century.

It is true that pre-colonial Africa was anything but an entirely homogenous entity. Some regions had developed sophisticated centralized states along lines quite similar to contemporary states, while others operated highly decentralized polities or what anthropologists refer to as stateless societies. Some pre-colonial African societies were matrilineal while others were patrilineal. Despite the differences, it is true that all pre-colonial African societies had a lot in common. In Chapter Five I noted that throughout pre-colonial Africa, without any exception, land was never treated as a commodity. Rather, it was communally held, with people being granted no more than use rights over the land. In Chapter Eight, I drew attention to the fact that throughout pre-colonial Africa, the same resource mobilization strategies, ranging from rotating credit associations to hometown associations were in operation. Chapter Nine drew attention to the importance of traditional healing strategies, which were remarkably similar throughout pre-colonial Africa. Finally, Chapter Ten revealed that geography and geology determined the nature of architecture and building techniques throughout pre-colonial Africa.

Amazingly, despite the virulent forces of colonialism and neocolonialism, most of the commonly shared traditions have survived to date. Some of the ancient commercial linkages and trade routes remain operational despite the existence of colonial borders that forcibly tore up and separated the indigenous communities. Cross-border trading, currently classified as smuggling, contributes untold sums to the GNPs of African countries today. This is despite the fact that hardly any efforts have been made to develop an international road network on the continent. No intended or unintended barrier, including those of the linguistic, political, social, economic or political genre, has succeeded in stopping interaction among Africans. International development authorities will do well to closely study and incorporate these informal interactions, which are occurring throughout the continent, into efforts to promote regional and even continent-wide integration.

Conclusion

This chapter underscores the importance of indigenous knowledge in the development planning process. The aim of underscoring the importance of indigenous knowledge (IK) is not to propose IK as a substitute for received knowledge – that is, knowledge borrowed from other cultures in all contexts. Rather, the aim is to advocate the use of IK as a complement to received knowledge. Perhaps more importantly, development planners and others involved in the development process in African countries must decide on which specific knowledge to use based on its comparative advantage in any given situation.

Central to the discussion in this chapter is the assumption that culture matters in the development process. Thus, planners and policymakers must bear in mind that unless development plans are based on indigenous African culture and values, they are likely to fail. Indigenous African cultural values and systems must constitute the basis for evaluating the appropriateness of received knowledge. The discussion in this chapter suggests that those involved in efforts to promote development in Africa will be remiss if they fail to incorporate indigenous knowledge in their repertoire of remedies to development problems on the continent. The discussion especially reveals that while African indigenous knowledge may appear deceptively mundane, it possesses a demonstrated capacity to tackle issues of critical importance to contemporary development efforts on the continent. In this connection, the chapter discussed ways in which African indigenous traditional practices and culture can facilitate the accomplishment of objectives such as regional politico-economic integration that have far-reaching implications for the continent's efforts to maximize its benefits from contemporary globalisation processes.

Bibliography

Adotevi, S. (1997). 'Cultural Dimensions of Economic Development and Political Integration in Africa,' In R. Lavergne (ed.), *Regional Integration and Cooperation in West Africa: A Multidimensional Perspective.* Ottawa/New Jersey: IDRC/Africa World Press.

African Ceremonies (On-Line). Photos of Sacred Rituals in Tribal Cultures. www.africanceremonies.com/ceremonies/photogallery.html

Africa-Union.org. The Official Site of the African Union: http://www.africa-union.org.

Afrol News (online). http://www.afrol.com/Categories/Women/FGM/

Akyeampong, E. (1997). 'Sexuality and Prostitution among the Akan of the Gold Coast c. 1650–1950.' *Past and Present*, 156, 144–173.

All Africa (Online). http:www.AllAfrica.com, posted on the web on 24 March 2005; retrieved, March 26, 2005.

Allman, J. (1996). 'Rounding up Spinsters: Gender Chaos and Unmarried Women in Colonial Asante.' *The Journal of African History*, 37 (2), 195–214.

Amadiume, I. (Online). 'Women and Development in Africa.' Feature Article on the Soka Gakkai International (SGI) website: http://www.sgi.org/english/Features/quarterly/0501/feature3.htm.

Amadiume, I. (1987). *Male Daughters and Female Husbands: Gender and Sex in an African Society.* London/New Jersey: Zed Books.

Amadiume, I. (1997). *Reinventing Africa: Matriarchy, Religion and Culture.* London: Zed Books.

Amos, L. (Online). 'Sankofa – Searching the Past to Connect to the Future.' Pittsburgh Teachers Institute (PTI). Curriculum Units online. Series: The Essentials of African Culture. www.chatham.edu.

Anderson, R. T. (1971). 'Voluntary Associations in History.' *American Anthropologist*, New Series, 73 (1), 209–222.

Anderson, D. and Rathbone, R. (eds.) (2000). *Africa's Urban Past.* Oxford/Portsmouth, NH: James Currey/Heinemann.

Andrews, P. A. (1971). 'Tents of the Tekna, Southwest Morocco,' in Oliver, P. (ed.), *Shelter in Africa.* New York/Washington: Praeger Publishers, pp. 124–142

Aprah, K. K. (2001). 'Culture, the Missing Link in Development Planning in Africa.' Paper Presented at the Roundtable Discussion on Mainstreaming Human Security and Conflict Issues in Long-Term Development Planning in Africa: A New Development Paradigm? Accra, 9–10 July 2001.

Apusigah, A. A. (2005). 'Re/inserting Performance in Social Constructivist Framing of Gender Meanings: Appreciating Female Sexuality in a Ghanaian Society.' Paper Presented at the Writing African Women – Poetics and Politics of African Gender Research Conference. Held at the University of the Western Cape, Cape Town, South Africa, 19–22 January 2005.

Ardener, S. (1953). 'The Social and Economic Significance of the Contribution Club among a Section of the Southern Ibo.' In Proceedings, Annual Conference, West African Institute of Social and Economic Research. Ibadan: West African Institute of Social and Economic Research.

Ardener, S. (1953). "The Social and Economic Significance of the Contribution Club among a Section of the Southern Ibo," Proceedings of the Annual Conference, West African Institute of Social and Economic Research (Sociology Section), Ibadan, pp. 128–142.

Ardener, S. (1964). 'The Comparative Study of Rotating Credit Associations.' *The Journal of the Royal Anthropological Institute of Great Britain and Ireland*, 94 (2), 201–229.

Arndt, S. (2000). 'African Gender Trouble and African Womanism: An Interview with Chikwenye Ogunyemi and Wanjira Muthoni.' *Signs*. 25 (3), 709–726.

Arnold, S. (1990) *Culture and Development in Africa.* Trenton, NJ: Africa World Press.

Asojo, A. (Online). 'Traditional African Architecture and its Impact on Place-Making: Case Studies from African and African-American Communities.' Paper presented at the 'Places of Cultural Memory: African Reflections on the American Landscape' Conference organized by the US Dept. of the Interior and held in Atlanta, GA, May 9–12, 2001. Available On-Line at http://www.cr.nps.gov/crd/Conferences/AFR_127-134_Asojo.pdf.

Atkinson, D. (Online). 'Vitex Agnus Castus: A Review.' *Positive Health: Complementary Health Magazine.* Online at: http://www.positivehealth.com (Retrieved: March 21, 2006).

Atkinson, G. A. (1950). 'African Housing.' *African Affairs*, 49 (196), pp. 228–237.

Austen, R. A. and Headrick, D. (1983). 'The Role of Technology in the African Past.' *African Studies Review*, 26 (3/4), 163–184.

Awe, B. (1991). 'Writing Women into History: The Nigerian Experience.' In Offen, K. et al. (eds.), *Writing Women's History: International Perspectives.* Bloomington, IN: Indiana University Press.

Baron, D. (1978). *Land Reform in Sub-Saharan Africa: An Annotated Bibliography.* Washington, DC: US Agency for International Development.

Barrows, R. and M. Roth (1990). 'Land Tenure and Investment in African Agriculture: Theory and Evidence.' *The Journal of Modern African Studies*, 28, 265–97.

Bascom, W. R. (1952). 'The Esusu: A Credit Institution of the Yoruba.' *Journal of the Royal Anthropological Institute*, 82, 63–69.

Battan, J. F. (1999). 'The "Rights" of Husbands and the "Duties" of Wives: Power and Desire in the American Bedroom, 1850–1910.' *Journal of Family History*, 24, 165–86.

Bauer, P. T. (1972). *Dissent on Development: Studies and Debates in Development Economics.* Cambridge, MA: Harvard University Press.

BBC (On-Line). 'Indian handheld to tackle digital divide.' On-line news item reported by correspondent, Mark Ward (retrieved, on March 25, 2005), http://www.news.bbc.co.uk.

Beall, J. (2002). 'Globalization and social exclusion in cities: framing the debate with lessons from Africa and Asia.' *Environment & Urbanization,* 14 (1), 41–51.

Bennett, T. W. (1998). 'Using Children in Armed Conflict: A Legitimate African Tradition?' Monograph No. 32. Available electronically at: http://www.iss.co.za/Pubs/Monographs/No32/UsingChildren.html. Retrieved on 10 December 10, 2005.

Berger, I. and White, E. F. (1990). *Women in Sub-Saharan Africa.* Soka Gakkai International (SGI), No. 39, Online Publication, http://www.sgi.org/english/Features/quarterly/0501/feature3.htm

Berman, B; Eyoh, D. and Kymlicka, W. (eds.) (2004). *Ethnicity and Democracy in Africa.* Oxford: James Currey.

Besley, T. (1995). 'Nonmarket Institutions for Credit and Risk Sharing in Low-Income Countries.' *The Journal of Economic Perspectives,* 9 (3), 115–127.

Besteman, C. (1994). 'Individualisation and the Assault on Customary Tenure in Africa: Title Registration Programmes and the Case of Somalia.' *Africa: Journal of International African Institute,* 64 (4): 484–515.

Binns, T. and Nel, E. (1999). 'Beyond the Development Impasse: The Role of Local Economic Development and Community Self-Reliance in Rural South Africa.' *The Journal of Modern African Studies,* 37 (3), 389–408.

Black College (Online) 'Black College/University Student: A History of Education Book II.' Available electronically. http://community-2.webtv.net/NUBIA/BLACKWORLDNUBIAN/.

Blier, S. P. (1983). 'Houses are Human: Architectural Self-Images of Africa's Tamberma.' *The Journal of the Society of Architectural Historians,* 42 (4), 371–382.

Bloomer, Kent and Moore, Charles (1977). *Body, Memory and Architecture.* New Haven: Yale University Press.

Bourdier, J-P. and Minh-ha, T. (1985). *African Spaces: Designs for Living in Upper Volta.* Teaneck, NJ: Holmes and Meier Publishers.

Boserup, E. (1970). *Women's Role in Economic Development.* London: Allen and Unwin.

Bosman, W. (1705). *A New and Accurate Description of the Coast of Guinea, Divided into the Gold, the Slave, and the Ivory Coasts.* London: J. Knapton.

Bridges.Org (On-Line). 'Evaluation of the SATELLIFE PDA project, 2002: Testing the use of handheld computers for healthcare in Ghana, Uganda and Kenya.' (retrieved, 25 March 2005) http://www.bridges.og/stellife/index.html.

Burgess, R. (1978). 'Petty commodity housing or dweller control? A critique of John Turner's view on housing policy.' World Development. 6 (9/10), pp. 1105–1133.

Bwakali, D. J. (2001). 'Gender Inequality in Africa.' *Contemporary Review*, an Online publication accessible through: http://www.findarticles.com.

Chatelain, H. (1889). 'Angolan Customs.' *The Journal of American Folklore*, 9 (32), 13–18.

Chavunduka, G. (Online). 'Christianity, African Religion and African Medicine.' Unpublished Paper Available on the Website of the World Council of Churches at: http://www.wcc-coe.org/wcc/what/interreligious/cd33-02.html.

Chilver, E. M., (1966). *Zintgraff's Explorations in Bamenda and the Benue Lands 1889–1892* , Buea, Cameroon: Ministry of Primary Education & Social Welfare and W. Cameroon Antiquities Commission Volume 2. (Previously circulated privately in 1961).

Chilver, E. M., (1967a). 'Paramountcy and protection in the Cameroons: the Bali and the Germans, 1889–1913', In: P. Gifford, and W.R. Louis (eds.) *Britain and Germany in Africa: Imperial Rivalry and Colonial Rule.* New Haven: Yale University Press.

Chilver, E. M., (1989) 'Women Cultivators, Cows and Cash-Crops: Phyllis Kaberry's Women of the Grassfields Revisited', In: P. Geschiere, and P. Konings (eds.) *Proceedings/Contributions, Conference on the Political Economy of Cameroon – Historical Perspectives, June 1988 (African Studies Centre Leiden, Research Reports No. 35)* : 383–422, Leiden: African Studies Centre.

Chilver, E. M., and P. M. Kaberry (1963). 'Traditional Government in Bafut, West Cameroon', *Nigerian Field* , 28, 1: 4–30.

Chilver, E. M. and Kaberry, P. M. (1966). *Notes on the Precolonial History and Ethnography of the Bamenda Grassfields* (mimeographed). Reissued 1968 in a revised form under the title *Traditional Bamenda,* Buea; Government Printing Press.

Chilver, E. M., and P. M. Kaberry (1968) *Traditional Bamenda: The Pre-colonial History and Ethnography of the Bamenda Grassfields*, Buea, Cameroon: Ministry of Primary Education & Social Welfare and West Cameroon Antiquities Commission, Volume 1.

Chojnacka, H. (1980), 'Polygyny and the Rate of Population Growth.' *Population Studies*, 34: 91–107.

Cipriani, L. (1938). *La Abitazioni Indigeni del A.O.J.*, Milano, pp. 47–50.

Clegg, N. (Online). 'Herbal Medicine, Antibiotics and the Imune System.' *Positive Health: Complementary Health Magazine.* Online at: http://www.positive health. com (Retrieved 21 March 2006).

Clignet, R. (1970). *Many Wives, Many Powers: Authority and Power in Polygynous Families.* Evanston, IL: Northwestern University Press.

Cohen, J. M. (1974). 'Peasants and Feudalism in Africa: The Case of Ethiopia.' *Canadian Journal of African Studies*, 8 (10: 155–157.

Cole, P. M. (2004). 'Cultural Competence now Mainstream Medicine: Responding to Increasing Diversity and Changing Demographics.' *Postgraduate Medicine*, 116 (6): 51–53.

Coquery-Vidrovitch, C. (1991). 'The Process of Urbanization in West Africa: From Origins to the Beginning of Independence.' *African Studies Review*, 34 (1), 1-98.

Coreil, J., Bryant, C. A. and Henderson, J. N. (2001). *Social and Behavioral Foundations of Public Health*. London: Sage.

Cornborough, J. (Online). 'Neem: An Ancient Cure for a Modern World.' *Positive Health: Complementary Health Magazine*. Online at: http://www.positive health. com (Retrieved 21 March 2006).

Courtright, P.; Chirambo, M.; Lewallen, S.; Chana, H.; and Kanjaloti, S. (2000). *Collaboration with African Traditional Healers for the Prevention of Blindness*. Singapore/New Jersey/London/Hong Kong: World Scientific.

Creative Exchange (Online). 'Highlights and Summary: Is Culture a Hidden Asset of Development – Implicit yet Invisible?' A publication of Creative Exchange, the Network for Culture and Development. Available online at: http://www. healthcomms.org/pdf/c-exch-culture.pdf.

Crisman, P. (Online). 'Style.' An article written for Whole Building Design Guide, found at http://wbdg.org, Retrieved on 10 April 2006.

Crook, R. C. (1986). 'Decolonization, the State, and Chieftancy in the Gold Coast.' *African Affairs*, 85 (338), 75–105.

Dalgleish, D. (2005). 'Pre-Colonial Criminal Justice in West Africa: Eurocentric Thought Versus Africentric Evidence. *African Journal of Criminology and Justice Studies*, Vol. 1 (1), 56–69.

Dalvi, S. (Online). 'Sutherlandia as Support for Immune Dysfunction.' *Positive Health: Complementary Health Magazine*. Online at: http://www.positive health. com (Retrieved: March 21, 2006).

Davidson, B. (1959). *Lost Cities of Africa*. Boston, MA: Little Brown and Co.

Davidson, B. (1970). *The Lost Cities of Africa*. Boston: Little Brown and Co.

Davidson, B. (1974). *Can Africa service argument against growth without development?* Boston/Toronto: Little, Brown & Company.

Davidson, J. (1996). *Voices from Mutira: Change in the Lives of Rural Gikuyu Women, 1910–1995*. 2nd edition. Boulder, CO: Lynne Rienner.

Davies, R. (Online). 'The Incredible Journey of Herbal Medicine.' *Positive Health: Complementary Health Magazine*. Online at: http://www.positive health.com (Retrieved: March 21, 2006).

Delafosse, M. (1911). 'Memorandum on Land Tenure in French West Africa.' *Journal of the Royal African Society*, 10 (39): 258–273.

Delafosse, M. (1911). 'Memorandum on Land Tenure in French West Africa.' *African Affairs*, 10, 258–273.

DeLancey, M. W. (1977). 'Credit for the Common Man in Cameroon.' *The Journal of Modern African Studies*, 15 (2), 316–322.

DeLancey, M. W. (1987). 'Women's Cooperatives in Cameroon: The Cooperative Experiences of the Northwest and Southwest Provinces.' *African Studies Review*, 30 (1), 1–18.

Denyer, S. (1978). *African Traditional Architecture: An Historical and Geographical Perspective*. London: Heinemann/New York: Africana Publishing Company.

Dillon, R. G. (1977). 'Ritual, Conflict, and Meaning in an African Society.' *Ethos*, 5 (2), 151–173.

Dillon, R. G., (1973). *Ideology, Process and Change in pre-colonial Meta*, Philadelphia: Unpublished PhD; University of Pennsylvania; Philadelphia.

Doebele, W. A. (1987). 'Land Policy.' In L. Rodwin (ed.), *Shelter, Settlement and Development*. Boston: Allen & Unwin.

Dorjan, V. R. (1959). 'The Factor of Polygyny in African Demography,' in W.R. Bascom and M.J. Herskovits (eds.), *Continuity and Change in African Cultures*. Chicago, IL: University of Chicago Press.

Drummond-Hay, J. C. (1925) 'An Assessment Report on the Clans of the Ndop Area in the Bamenda Division of the Cameroons Province', *Cameroon National Archives*, Buea (E.P. 1282).

Duignan, P. and Gann, L. H. (1975). 'The pre-colonial economies of sub-Saharan Africa.' in Duignan, P. and Gann, L. H. (eds.), *Colonialism in Africa 1870–1960, Vol. 4: The Economics of Colonialism*. London/New York/Melbourne: Oxford University Press.

Edeinya, I. (1970). 'Anti-Malarial activity of Nigerian Neem Leaves.' *Trans Royal Soc Tropical Medicine*, 87 (4): 471.

Eyong, S. O. (1990). 'Traditional Housing in African Cities.' Review of Schwerdtfeger, F.W., (1982). *Traditional Housing in African Cities*. John Wiley, Brighton. *Africa: Journal of the International African Institute*, 60 (2), 304–306.

Falade, J. B. (1990). 'Yoruba Palace Gardens.' *Garden History*, 18 (1), 47–56.

Fall, Y. (1997). 'Gender Relations in the Democratization Process: An Analysis of Agrarian Policies in Africa.' *Issue: A Journal of Opinion*, 25 (2): 8–11.

FAO (On-Line). 'Interview with Mr. Paolo Groppo-February 2003.' www.rdfs.net/new/interviews/0302in/0302in_gropo_en.htm

Feder, G. and R. Noronha (1987). 'Land Rights Systems and Agricultural Development in Sub-Saharan Africa.' *Research Observer*, 2, 142–71.

Feeney, D. (1982). *The Political Economy of Productivity: Thai Agricultural Development 1880–1875*. Vancouver: University of the British Columbia.

Feierman, S. (1985). 'Struggles for Control: The Social Roots of Health and Healing in Modern Africa.' *African Studies Review*, 28 (2/3), 73-147.

Ferguson, W. J. and Candib, L. M. (2002). 'Culture, Language, and Doctor-Patient Relationship.' *Family Medicine*, 34 (5), 353–361.

Flint, J. E. (1966). *Nigeria and Ghana*. Englewood Cliffs/New Jersey: Prentice-Hall, Inc.

Frascari, M. (1991). *Monsters of Architecture: Anthropomorphism in Architectural Theory*. New York: Rowman & Littlefield Publishing.

Fraser, D. (1968). *Village Planning in the Primitive World*. New York: Braziller.

Frazer, J. G. and Downie, R. A. (1938). *The Native Races of Africa and Madagascar: Anthologia Anthropologica*. London: Routledge.

Gardi, R. (1973). *Indigenous African Architecture*. (Engl. Trans.: S. MacRae). New York: Van Nostrand Reinhold.

Gebremedhin, N. (1971). 'Some Traditional Types of Housing in Ethiopia.' In Oliver, P. (ed.) (1971). *Shelter in Africa.* New York: Barrie & Jenkins.

Geertz, C. (1962). 'The Rotating Credit Association: A 'Middle Rung' in Development.' *Economic Development and Cultural Change*, 10 (3), 241–263.

Geschiere, P. (1993). 'Chiefs and Colonial Rule in Cameroon: Inventing Chieftaincy, French and British Style.' *Africa: Journal of the International African Institute*, 63 (2), 151–175.

Giblin, J. (2005). *A History of the Excluded: Making Family and Memory a Refuge from State in Twentieth-Century Tanzania.* Oxford: James Currey.

Giblin, J. (Online). 'Introduction: Diffusion and other Problems in the History of African States.' Available on the website, Art & Life in Africa: http://www.uiowa.edu/~africart/toc/people/Asante.html.

Giddens, A. (1984). *The Nation-State and Violence.* Berkeley, CA: University of California Press.

Gildea, Jr., R. Y. (1964). 'Culture and Land Tenure in Ghana.' *Land Economics*, 40 (1): 102–104.

Goheen, M. (1988). 'Land Accumulation and Local Control: The Manipulation of Symbols of Power in Nso, Cameroon,' in Downs, R.E. and Reyna, S.P. (eds.) *Land and Society in Contemporary Africa.* Hanover: University Press of N. England.

Goheen, M. (1995). 'Gender and Accumulation in Nso',' Paideuma 41, 1995, pp. 73–81.

Goheen, M. (1996). 'Accidental Collision: A Conversation between Sally Chilver, Mitzi Goheen and Eugenia Shanklin,' *Journal of the Anthropological Society of Oxford.*

Goose, D. H. (1963). 'Toot-Mutilation in West Africans.' *Man*, 63, 91–93.

Gordon, A. (1996). *Transforming Capitalism and Patriarchy: Gender and Development in Africa.* London: Lynne Rienner.

Gray, L. and Kevane, M. (1999). 'Diminished Access, Diverted Exclusion: Women and Land Tenure in Sub-Sahara Africa.' *African Studies Review*, 42 (2): 15–39.

Gray, R. (1982). 'Christianity, Colonialism, and Communication in Sub-Saharan Africa.' *Journal of Black Studies*, 13 (1), 59–72.

Gregg, C. J. A. (1924). 'Meta: An Assessment Report on the Meta Clan of the Bamenda Division, Cameroons Province.' Unpublished Official Document of the Colonial Government of British Southern Cameroons.

Griffiths, I. L. (1995) *The African Inheritance.* London/New York: Routledge.

Groneman, C. (1994) 'Nymphomania: The Historical Construction of Female Sexuality,' *Signs*, 19 (2), 337–367.

Gutkind, E. A. (1953). 'How Other Peoples Dwell and Build – Indigenous Houses of Africa.' *Architectural Design*, 23, 121–124.

Gyasi, E. A. (1994). 'The Adaptability of African Communal Land Tenure to Economic Opportunity: The Example of Land Acquisition for Oil Palm Farming in Ghana.' *Africa: Journal of the International African Institute*, 64 (3): 391–405.

Gyekye, K. (1996). *African Cultural Values: An Introduction*. Philadelphia: Sankofa Publishing Co.

Gyekye, K. (1997). *Tradition and Modernity: Philosophical Reflections on the African Experience*. New York: Oxford University Press.

Haley, H. (1976). *Roots: The Saga of an American Family*. New York: Dell Publishing

Halfani, M. (1996). 'Marginality and dynamism: Prospects for the sub-Saharan African city,' in Cohen, M., Ruble, B., Tulchin, J. and Garland, A. (eds.), *Preparing for the urban future: Global pressures and local forces*. Washington, DC.: Woodrow Wilson Center Press.

Hamilton, R. W. (1920). 'Land Tenure Among the Bantu Wanyika of East Africa.' *Journal of the Royal African Society*, 20 (77): 13–18.

Hanson, H. E. (2003). *Landed Obligation: The Practice of Power in Buganda*. Portsmouth: Heinemann.

Hastings, A. (1997). *The Construction of Nationhood: Ethnicity, Religion and Nationalism*. Cambridge University Press.

Hegel, G. W. F. (1956). *The Philosophy of History*. New York: Dover.

Hegel, G. W. F. (1991). *The Philosophy of History*. Trans.: J. Sibree (Great Books in Philosophy). New York: Prometheus Books.

Herbal Africa (Online). 'Natural Traditional African Medicines.' Available electronically at: http://www.herbalafrica.co.za

Herbst, J. (2000). *States and Power in Africa: Comparative Lessons in Authority and Control*. Princeton, NJ: Princeton University Press.

Herodotus, II (1891). *Cap. 84*, (Trans. Cary). London.

Herskovits, M. J. (1952). 'Some Problems of Land Tenure in Contemporary Africa.' *Land Economics*, 28 (1): 37–45.

Hodges, D. L. and S. McCurdy (1996). 'Wayward Wives, Misfit Mothers, and Disobedient Daughters: "Wicked" Women and the Reconfiguration of Gender in Africa.' *Canadian Journal of African Studies*, 30 (1), 1–9.

Honey, R. and Okafor, S. (1998). *Hometown Associations: Indigenous Knowledge and Development in Nigeria*. Rugby, UK: Intermediate Technology Development Group Publishing.

Horton, M. (1991). 'Primitive Islam and Architecture in East Africa.' *Muqarnas*, 8, 103–116.

Hughes, A. J. B. (1962). 'Some Swazi Views on Land Tenure.' *Africa: Journal of the International African Institute*, 32 (3): 253–258.

Hull, R. W. (1976). *African Cities and Towns Before the European Conquest*. New York/London: W.W. Norton & Company.

Hunt, N. R. (1988). '"Le Bebe en Brousse:" European Women, African Birth Spacing and Colonial Intervention in Breast Feeding in the Belgian Congo.' *The International Journal of African Historical Studies*, 21 (3), 401–432.

ICRW (2005). Property Ownership for Women Enriches, Empowers and Protects: Toward Achieving the Third Millennium Development Goal to Promote Gender

Equality and Empower Women. Unpublished Paper, International Center for Research on Women, Washington, DC.

Ifeka-Moller, C. (1974). 'White Power: Social-Structural Factors in Conversion to Christianity, Eastern Nigeria, 1921–1926.' *Canadian Journal of African Studies,* 8 (1), 55–72.

ISHR-WAC (Online). 'Factors Inhibiting Women's Rights in West Africa.' Available online at: http://www.ishr.org/sections-groups/wac/africanwomen.htm.

Isong, C.N. (1958). 'Modernisation of the Esusu Credit Society.' Conference Proceedings, Nigerian Institute of Social and Economic Research.

Janelle, D. G. and Beuthe, M. (1997). 'Globalization and research issues in transportation.' *Journal of Transport Geography,* 5 (3), 199–206.

Jarosz, L. (1992). 'Constructing the Dark Continent: Metaphor as Geographic Representation of Africa.' *Geografiska Annaler. Series B, Human Geography,* 74 (2), 105–115.

Jarrett, A. (1996). *The Under-Development of Africa: Colonialism, Neo-Colonialism and Socialism.* Lanham/New York/London: University Press of America.

Jenkins, P.; Robson, P.; and Cain, A. (2002). 'Local responses to globalization and peripheralization in Luanda, Angola.' *Environment & Urbanization,* 14 (1), 115–127.

Jinadu, L. A. (1978) 'Some African Theorists of Culture and Modernization: Fanon, Cabral and Some Others.' *African Studies Review,* 21 (1) 121–138.

Jones, W. F. (1907). 'Report on the Human Remains.' *Archaeological Survey of Nubia,* 2, 263.

Kaberry, P. (1952). *Women of the Grassfields.* (Colonial Research Publication No. 14). London: HMSO.

Kandawire, J. A. K. (1977). 'Thangata in Pre-Colonial and Colonial Systems of Land Tenure in Southern Malawi with Special Reference to Chingale.' *Africa: Journal of the International African Institute,* 47 (2): 185–191.

Kanyemba, J. (Online). 'Preserving Traditional Building Materials and Construction Methods by the Use of Performance Based Building Codes.' Proceedings of the International Conference on Creating a Sustainable Construction Industry in Developing Countries held at Stellenbosch, South Africa, 11–13, November 2002. Available online at: http://www.odsf.co.za/cdcproc/3rd_proceedings.html.

Katz, A. H. (1981). 'Self-Help and Mutual Aid: An Emerging Social Movement?' *Annual Review of Sociology,* 7, 129–155.

Kenyatta, J. (1938). *Facing Mount Kenya: The Tribal Life of the Gikuyu.* London: Secker and Warburg.

Kerr, G. B. (1978). 'Voluntary Associations in West Africa: 'Hidden' Agents of Social Change.' *African Studies Review,* 21 (3), 87–100.

Khan, M. and Wassilew, S. W. (1987). 'The Effect of Raw Material from Neem Extracts on Fungi Pathogenic to Humans'. In Shmutterer and Ascher. *Natural Pesticides from the Neem Tree and other Tropical Plants.* International Neem Conference. Nairobi, Kenya, pp. 645–650.

Khapoya, V. (1998). *The African Experience: An Introduction* (2nd ed.). Upper Saddle River, NJ: Prentice Hall.

King, A. D. (1990). *Urbanism, Colonialism and the World Economy: Cultural and Spatial Foundations of the World Urban System.* New York: Routledge.

King, L. D. (2001). 'State and Ethnicity in Precolonial Northern Nigeria.' *Journal of Asian and African Studies*, XXXVI, 4: 339–360.

Kingsley, M. H. (1897). 'The Fetish View of the Human Soul.' *Folklore*, 8 (2), 138–151.

Ki-Zerbo, J. (1978). *Histoire de l'Afrique Noire.* Paris: Hatier.

Klitz, W.A. (On-Line). 'The African Creation Story.' Retrieved: 21 November 2005. http://dickinsg.intrasun.tcnj.edu

Kouba, L. J. and Muasher, J. (1985). 'Female Circumcision in Africa: An Overview.' *African Studies Review*, 28 (1), 95–110.

Kpone-Tonwe, S. (2001). 'Leadership Training in Precolonial Nigeria: The Yaa Tradition of Ogoni.' *The International Journal of African Historical Studies*, 34 (2), 385–403.

Kuper, H. and Kaplan, S. (1944). 'Voluntary Associations in an Urban Township', *African Studies*, 3, 178-186.

Lacey, M. J. (2002). 'Foreword,' in Selznick, P., *The Communitarian Persuasion.* Washington, DC: Woodrow Wilson Center Press/Johns Hopkins University Press.

Leaky, L. (1977). *The Southern Kikuyu.* London: Academic Press, 1977.

Lebeuf, J-P. (1967). 'L'Architecture Africaine Traditionelle,' Coloque, 1er Festival Mondial des Arts Nègres, Dakar.

Le Corbusier (Jeanneret, Charles Edouard) (1926). *Vers une architecture* [Towards a New Architecture]. Trans. F. Etchells from the 13th French ed. 1960, (orig. pub.: 1923). Orlando, Fl: Holt, Rinehart and Winston.

Leith, J. (Online). 'Devil's Claw: Sustainable Harvesting of and Fair Trade Medicinal Plants.' *Positive Health: Complementary Health Magazine.* Online at: http://www.positive health.com (Retrieved: March 21, 2006).

Levine, N. and Sagree, W. H. (1980). 'Women with Many Husbands: Polyandrous Alliance and Marital Flexibility in Africa and Asia.' Special Issue of the Journal of Comparative Family Studies, XI (3) (entire issue).

Levtzion, N. and Spaulding, J. (eds) (2003). Medieval West Africa: Views from Arab Sholars and Merchants. Princeton, NJ: Markus Wiener Publishers.

Lewis, D. (On-Line). 'African Gender Research and Postcoloniality: Legacies and Challenges.' Paper available on-line at: http://www. Codesria.org/Links/conferences/gender/LEWIS.pdf.

Little, K. (1965). *West African Urbanization: A Study of Voluntary Associations in Social Change.* London: Cambridge University Press.

Livingston, T. W. (1974). 'Ashanti and Dahomean Architectural Bas-Reliefs.' *African Studies Review*, 17 (2), 435–448.

Lloyd, C. (2000). 'Globalization: Beyond the ultra-modernist narrative to a critical realist perspective on geopolitics in the cyber age.' *International Journal of Urban and Regional Research,* 24 (2), 258–73.

Macdonald, J. (1890). 'Manners, Superstitions, and Religions of South African Tribes.' The Journal of the Anthropological Institute of Great Britain and Ireland, 19, 264–296.

Maier, D. (1979). 'Nineteenth-Century Asante Medical Practices.' *Comparative Studies in Society and History,* 21 (1), 63-81.

Majoe, M. (Online). 'Kigelia Africana for Skin Conditions: Herbal Relief for Eczema and Psoriasis.' *Positive Health: Complementary Health Magazine.* Online at: http://www.positive health.com (Retrieved: March 21, 2006).

Mallet, V. (On-Line). 'The Seven-Day Week.' (Retrieved, 20 November, 2005). http://www.ac.wwu.edu

Mandivamba, R. (nd). 'Why Land Tenure is Central to Africa's Future Governance, Economic and Social Progress.' Unpublished paper available through the Scandinavian Seminar College: African Experiences of Policies and Practices Supporting Sustainable Development.

Mark, P. (1996). ' "Portuguese" Architecture and Luso-African Identity in Senegambia and Guinea, 1730–1890.' *History in Africa,* 23, 179–196.

Mathooko, J.M. (2000). 'The Status and Future of African Traditional Ecological Knowledge in the Sustainability of Aquatic Resources.' In the Proceedings of the 2nd Pan-African Symposium on the Sustainable Use of Natural Resources in Africa, held in Ouagadougou, Burkina Faso, 24–27 July 2000.

Mattingly, M. (1999). 'The role of government of urban areas in the creation of urban poverty.' In Jones, S. and Nelson, N. (eds.), *Urban poverty in Africa: From understanding to alleviation.* Southampton Row, London: Intermediate Technology Publication.

Maynard, K. (2004). *Making Kedjom Medicine: A History of Public Health and Well-Being in Cameroon.* Westport, CT/London: Praeger.

Mazrui, A. (1986). *The Africans: A Triple Heritage* (Video series), Washington, DC/London: WETA/BBC.

Mazrui, A. A. (1998). *The Africans: A Triple Heritage.* Boston/Toronto: Little, Brown & Company.

Mazrui, A. A. (2002). 'Globalization between the market and the military: A Third World perspective.' *Journal of Third World Studies,* XIX (1), 13–24.

McFadden, P. (2001). 'Cultural Practice as Gendered Excluion: Experiences from Southern Africa.' in *Discussing Women's Empowerment; Theory and Practice.* SIDA Studies No. 3.

McIntyre, A. (Online). 'Herbs – at the Forefront of Modern Medicine.' *Positive Health: Complementary Health Magazine.* Online at: http://www.positive health. com (Retrieved: March 21, 2006).

Meek, C. K. (1949). *Land Law and Custom in the Colonies.* Oxford: Oxford University Press.

Meillasoux, C. (1968). *Urbanization of an African Community: Voluntary Associations in Bamako*. Evanston, IL: Northwestern University Press.

Meredith, H. (1812). *An Account of the Gold Coast of Africa*. London: Frank Cass.

Mernissi, F. (1988). *Doing Daily Battle: Interviews with Moroccan Women*. London: Women's Press.

Metmuseum.org (On-line). 'Masterhand: Individuality and Creativity Among Yoruba Sculptors.' Retrieved on November 21, 2005 from www.metmuseum.org.

Michie, H. (Online) 'From Little Came a Lot: Africa's Greatness.' Essentials of African Culture (2004). Available electronically from the Pittsburgh Teachers Institute, Chatham College: http://www.chatham.edu/PTI.

Miracle, M. P.; Miracle, D.S.; and Cohen, L. (1980). 'Informal Savings Mobilization in Africa.' *Economic Development and Cultural Change*, 28 (4), 701–724.

Mufeme, E. (On-Line) 'Land: Breaking Bonds and Cementing Ties.' (Land and Spirituality. *Echoes*, http://www.wcc-coe.org/wcc/what/jpc/echoes-16-05.html.

Muhsam, H. V. (1956). 'The Fertility of Polygamous Marriages.' *Population Studies*, 10, 3–16.

Mulder, M.B. (1989). 'Marital Status and Reproductive Performance in Kipsigis Women: Re-Evaluating the Polygyny – Fertility Hypothesis.' *Population Studies*, 43: 285–304.

Myrdal, G. (1956). *An International Economy*. New York:

Nevin, Tom (2000). 'Africa and the globalisation conundrum' *African Business*, 258 (Oct.), pp. 8–10.

Njoh, A. (1995). 'Building and Urban Land Use Controls in Developing Countries: A Critical Appraisal of the Kumba (Cameroon) Zoning Ordinance.' *Third World Planning Review*, 17 (3), 337–356.

Njoh, A. (2000). 'Some Development Implications of Housing and Spatial Policies in Sub-Saharan African Countries with Emphasis on Cameroon.' *International Planning Studies*, 5 (1), 25–44.

Njoh, A. J. (1998). 'The political economy of urban land reforms in a post-colonial state.' *International Journal of Urban and Regional Research*, 22 (3), 408–24.

Njoh, A. J. (1999). *Urban Planning, Housing and Spatial Structures in Sub-Saharan Africa: Nature, Impact and Development Implications of Exogenous Forces*. Aldershot: Ashgate.

Njoh, A. J. (2003a). *Planning in contemporary Africa: The state, town planning and society in Cameroon*. Aldershot, UK: Ashgate.

Njoh, A. J. (2003b). 'The Role of Community Participation in Public Works Projects in LDCs: The Case of the Bonadikombo, Limbe (Cameroon) Self-Help Water Supply Project.' *International Development Planning Review*, 25(1), 85–103.

Njoh, A. J. (forthcoming). 'Determinants of Success of Self-Help Public Works Projects: The Case of the Kumbo Water Supply Project in Cameroon.' *International Development Planning Review*.

Nwabughuogu, A.I. (1984). 'The "Isusu:" An Institution for Capital Formation among the Ngwa Ibo; Its Origin and Development to 1951.' *Africa: Journal of the International African Institute*, 54 (4), 46–58.

O'Flaherty, M. (1998). 'Communal Tenure in Zimbabwe: Divergent Models of Collective Land Holding in the Communal Areas.' *Africa: Journal of the International African Institute,* 68 (4): 537–557.

Obijiofor, L. (n.d.). 'To Join or not to Join: Africa's Dilemma in the Age of Technology.' PDF file available on-line at http://www.wfsf.org/org/pub/publications/Brisbane_ 97/OBIJI.pdf. Retrieved, Dec. 10, 2005.

Ojameruaye, E. (2004). 'Strategies for Self-Reliant Economic Development of Urhoboland.' Paper presented at the 5th Annual Conference of Urhobo Historical Society at PTI Conference Centre, Delta State, Nigeria, October 30, 2004.

Ojo, G. J. A. (1966). *Yoruba Palaces.* London: University of London Press.

Okonjo, K. (1979). 'Rural Women's Credit Systems: A Nigerian Example.' *Studies in Family Planning,* 10 (11/12), 326–331.

Oliver, P. (ed.) (1971). *Shelter in Africa.* New York: Praeger.

Olusanya, P. O. (1971). 'The Problem of Multiple Causation in Population Analysis, with Particular Reference to the Polygamy – Fertility Hypothesis,' *Sociological Review,* 19, 165–178.

Optimist (1923). 'Missionaries and Education in Pagan Africa.' *Journal of the Royal African Society,* 23 (89), 44–47.

Ottenberg, S. (1968). 'The Development of Credit Associations in the Changing Economy of the Afikpo Ibo.' *Africa: Journal of he International African Institute,* 38 (3), 237–252.

Parpart, J. L. (2000). 'Gender and Colonial History.' Review of Hunt, N.R., Liu, T.P. and Quataert, J. (eds.) *Gendered Colonialisms in African History,* Blackwell, Oxford. *The Journal of African History,* 41 (1), 153–154.

Paulme, D. (1963). *Women of Tropical Africa.* (Trans. H.M. Wright), Berkeley: University of California Press.

Payne, B. (Online). 'Tisanes and their Use for Minor Ailments.' *Positive Health: Complementary Health Magazine.* Online at: http://www.positive health.com (Retrieved: 21 March 2006).

Pedersen, P. O. (2001). 'Freight transport under globalisation and its impact on Africa.' *Journal of Transport Geography,* 9, 85–99.

Perkin, J. (1993). *Victorian Women.* New York, : New York University Press.

Peterson, R.D.; Wunder, D.F.; and Mueller, H.L. (1999). *Social Problems: Globalization in the Twenty-First Century.* Upper Saddle River/New Jersey: Prentice Hall.

Place, F. and P. Hazell (1993). 'Productivity Effects of Indigenous Land Tenure Systems in Sub-Saharan Africa.' *American Journal of Agricultural Economics,* 75 (1): 10–19.

Pool, D. I. (1968) 'Conjugal Patterns in Ghana.' *Canadian Review of Sociology and Anthropology,* 5, 241–253.

Positive Health (Online). Positive Health Magazine: Integrated Medicine for the 21st Century. Online version of print British-based magazine for alternative medicine. http://www.positivehealth.com/

Power, J. (1995). '"Eating the Property": Gender Roles and Economic Change in Urban Malawi, Blantyre-Limbe, 1907–1953.' *Canadian Journal of African Studies*, 29 (1), 79–107.

Practical Action (Online). 'Traditional Communities' Indigenous Knowledge about Animal Breeding.' Practical Action: Technology Challenging Poverty. Available electronically at: http://practicalaction.org/?id=case_indigenous_knowledge.

Prussin, L. (1974). 'An Introduction to Indigenous African Architecture.' *The Journal of the Society of Architectural Historians*, 33 (3), 182–205.

Prussin, L. (1986). *Hatumere: Islamic Design in West Africa*. Berkeley/Los Angeles/London: University of California Press.

Prussin, L. (1999). 'Non-Western Sacred Sites: African Models.' *The Journal of the Society of Architectural Historians*, 58 (3), 424–433.

Purple Planet Medicine (Online). 'Real Magic: The Occult Library,' http://www.purpleplanet.co.uk.

Rasmussen, S.E. (1962). *Experiencing Architecture*. Cambridge, MA: Massachusetts Institute of Technology Press.

Reid, A.; Lane, P.; Segobye, A.; Borjeson, L.; Mathibidi, N. and Sekgarametso, P. (1997), *World Archaeology*, 28 (3), 307–392.

Reid, R. (2005). 'Nationhood, Power and History: Unfinished Business and the Longue Duree in Uganda,' Review of G. Thompson, *Governing Uganda: British Colonial Rule and its Legacy*. (Fountain Publishers, Kampala, 2003); and H.E. Hanson, *Landed Obligation: The Practice of Power in Buganda*. (2003) Portsmouth NH: Heinemann. *The Journal of African History*, 46: 321–325.

Religious Tolerance (Online) Religious Tolerance.Org. http://www.religioustolerance.org.

Reyna, S. P. and Bouquet, C. (1975) 'Chad.' In J. C. Caldwell (ed.), *Population Growth and Socio-economic Change in West Africa*. New York: Population Council.

Rodney, W. (1981). *How Europe underdeveloped Africa*. Washington, D.C.: Howard University Press.

Rostow, W. W. (1960). *The Stages of Economic Growth: A Non-Communist Manifesto*. Cambridge: Cambridge University Press.

Rowlands, M. (1993). 'Accumulation and the Cultural Politics of Identity in the Grassfields.' In Geschiere, P. and Konings, Piet (eds.), *Itinéraires d'Accumulation au Cameroun: Pathways to Accumulation in Cameroon*. Paris: Karthala, pp. 71–97.

Ruffer, M. (1921). *Studies in Paleopathology of Egypt*. Chicago: University of Chicago Press.

Rutman, G. L. (1969). 'Innovation in the Land Tenure System of the Transkei, South Africa.' *Land Economics*, 45 (4), 467–471.

Sacks, K. (1974). 'Engels Revisted: Women, the organization of production, and private property,' In Rosaldo and Lamphere, *Women, Culture and Society*, Standford: Stanford University Press, pp. 206–222.

Sassen, S. (2002). 'Locating cities on global circuits.' *Environment & Urbanization*, 14 (1), 13–30.

Schmidt, E. (1992). *Peasants, Traders, and Wives: Shona Women in the History of Zimbabwe, 1870–1939*. Portsmouth, NH: Heinemann.

Schneider, G. E. (2003). 'Globalization and the poorest of the poor: Global integration and the development process in sub-Saharan Africa.' *Journal of Economic Issues*, XXXVII (2), 389–96.

SciDev.Net (On-Line). 'Internet Brings U.S. Experiments to African Students.'

Schwerdtfeger, F. (1971). 'Housing in Zaria.' In Oliver, P. (ed.), *Shelter in Africa*. New York/Washington: Praeger Publishers, pp. 59–79.

Scott, G. (1999). *The Architecture of Humanism: A Study in the History of Taste.* (Orig. pub., 1914). New York: W.W. Norton & Company.

Scruton, R. (1979). *The Aesthetics of Architecture*. London: Methuen and Co. Ltd.

Seers, D. (1969). 'The Meaning of Development.' *International Development Review*, 3.

Seers, D. (1977). 'The New Meaning of Development.' *International Development Review*, 3.

Seidman, A. and Seidman, R. B. (1984). 'The Political Economy of Customary Law in the Former British Territories of Africa.' *Journal of African Law*, 28 (1/2): 44–55.

Selznick, P. (2002). *The Communitarian Persuasion*. Washington, DC: Woodrow Wilson Center Press/Johns Hopkins University Press.

Sen, A. (2004). 'How does Culture Matter?' In V. Rao and M. Walton (eds.), *Culture and Public Action*. Stanford, CA: Stanford University Press.

Shafritz, J. M. and Russell, E. W. (1996). *Introducing Public Administration*. New York: Longman.

Shaw, C. M. (1995). *Colonial Inscriptions: Race, Sex, and Class in Kenya*. Minneapolis, MN: University of Minnesota Press.

Shipton, P. and Goheen, M. (1992). 'Introduction. Understanding African Land-Holding: Power, Wealth, and Meaning.' *Africa: Journal of the International Institute*, 62 (3): 307–325.

Shipton, Parker (1994). 'Land and Culture in Tropical Africa: Soils, Symbols, and the Metaphysics of the Mundane.' *Annual Review of Anthropology*, 23: 347–377.

Shuja, Sharif M (2001) 'Coping with globalisation' *Contemporary Review*, 279 (1630) (Nov), 257–63.

Smeldy, G. D.; Stith, A. Y. and Nelson, A. R. (eds.) (2003). *Unequal Treatment: Confronting Racial and Ethnic Disparities in Health Care*. Washington, DC: National Academies Press.

Smiley, E. (Online). 'The New Traditional Family.' Political Affairs.Net: Marxist Thought Online. At http://www.politicalaffairs.net/article/articleview/111/1/29, Retrieved, March 28, 2006.

Smith, A. (1986). *The Ethnic Origins of Nations*. Oxford: Blackwell.

Smith, E. G. (1908). 'The Most Ancient Splints.' *British Medical Journal*, I, 732–734.

Sorenson, J. (2003). (ed.) *Disaster and Development on the Horn of Africa*. London: Macmillan.

Sorrenson, M. P. K. (1967). *Land Reform in the Kikuyu Country: A Study of Government Policy*. Oxford: Oxford University Press.

Southall, A. W. and Gutkind, C. W. (1956). 'Townsmen in the Making: Kampala and its Suburbs.' Kampala: *East African Institute of Social Research. 9.*

Stanley, H. M. (1876). *Through the Dark Continent*, Vol. I. Harper Bros.

Steady, F. C. (1981). (ed.), *The Black Woman Cross-Culturally* Cambridge: Sekenkman.

Stennard, L. (Online). 'Devil's Claw: Therapeutic Uses.' *Positive Health: Complementary Health Magazine.* Online at: http://www.positive health.com (Retrieved: March 21, 2006).

Stolberg, S. G. (2002). 'Race Gap Seen in Health Care of Equally Insured Patients.' *New York Times*. (March 21).

Stopford, J.G.B. (1901). 'Glimpses of Native Law in West Africa.' *Journal of the Royal African Society*, 1 (1), 80–97.

Strobel, M. (1982). 'African Women.' *Signs*, 8 (1), 109–131.

Sudarkasa, N. (1986). 'The Status of Women in Indigenous African Societies.' Feminist Studies, 12 (1), 91–103.

Tembo, M. S. (Online). 'Traditional Patterns in Africa.' Electronic document available at: http://www.Bridgewater.edu/~mtembo/africantraditionalfamily.htm.

Thomas, B. P. (1987). 'Development Through Harambee: Who Wins and Who Loses?: Rural Self-Help Projects in Kenya.' *World Development*, 15 (4), 463–481.

Thomas, N. W. (1913). *Anthropological Report on the Ibo-Speaking Peoples, Part I, Law and Customs of the Ibo of the Awka Neighbourhood, Southern Nigeria.* London: Harrison.

Thomas, N.W. (1916). *Anthropological Report on Sierra Leone, Part I, Law and Customs of the Time and other Tribes.* London: Harrison.

Thompson, G. (2003). *Governing Uganda: British Colonial Rule and its Legacy.* Kampala: Fountain Publishers.

Todd, T. W. (1921). 'Egyptian Medicine: A Critical Study of Recent Evidence.' *American Anthropologist, New Series*, 23 (4), 460–470.

Tomasugi, T. (1980). *A Structural Analysis of Thai Economic Theory: A Case Study of a Northern Chao Phraya Delta Village.* Tokyo: Institute for Economics.

Tordoff, W. (1984). *Government and Politics in Africa.* London: Macmillan.

Toynbee, A. (1934). *A Study of History: Introduction, The Genesis of Civilization.* Oxford: Oxford University Press.

Turner, S. (Online). 'The Reality of the Golden Age of West African Civilization.' Pittsburgh Teachers Institute (PTI) Curriculum Units Online. Series: The Essentials of African Culture.www.chatham.edu.

Tutu II, O. (2004). 'Traditional Systems of Governance and the Modern State: Keynote Address Presented by His Royal Majesty Otumfuo Osei Tutu II, Asantehene at the Fourth African Development Forum,' Addis Ababa, October 12, 2004.

Universal House of Justice (Online) 'Baha'i Topics: An Information Resource'. Available at: http://info.bahai.org.

US Census Bureau (Online). 'Households, by Type: 1940 to Present.' See Table HH-1, U.S. Census Bureau Internet Release Date: May 25, 2006. http://ask.census.gov.

Van den Brink, R. and Chavas, J-P. (1997). 'Microeconomics of an Indigenous African Institution: The Rotating Savings and Credit Association.' *Economic Development and Cultural Change*, 45 (4), 745–772.

Varma, P. (2000). 'Technical and vocational education and development.' *Development Express*. (Canadian International Development Agency), No. 4 (1999–2000).

Versi, A. (2000). 'Globalisation and Africa.' *African Business*, No. 258, p. 7.

Vicinus, M. (ed.) (1977), *A Widening Sphere: Changing Roles of Victorian Women*. Bloomington, IN: Indiana University Press.

Von Klein, C. H. (1905). 'The Medical Features of the Papyrus Ebers.' *Journal of American Medical Association*, 45, 1928–1935.

Warnier, J-P. (1985). *Echanges, Developpement et Hiérarchies dans le Bamenda Pré-colonial*, Wiesbaden, Stuttgart: Franz Steiner Verlag.

Warren, D. M. (1992). 'Indigenous knowledge, biodiversity conservation and development.' Keynote address at the International Conference on Conservation of Biodiversity in Africa: Local Initiatives and Institutional Roles, 30 August–3 September 1992, Nairobi, Kenya.

Waterston, A. (1965). *Development Planning: Lessons of Experience*. Baltimore, MD: John Hopkins University Press.

WCCD (On-Line). *Our Creative Diversity*. United Nations Educational, Scientific and Cultural Organization. Report by World Commission on Culture and Development, available on-line at: http://www.unesco.org/culture/policies/ocd/index.shtml.

Weber, M. (1905), *The Protestant Ethic and the Spirit of Capitalism*. New York: Scribner's Press, 1958 (trans. from the original German version of 1905).

White, L. (1990). *The Comforts of Home: Prostitution in Colonial Nairobi*. Chicago: University of Chicago Press.

WHO (2002). WHO Traditional Medicine Strategy 2002–2005. Geneva, WHO, 2002 (Document Reference No.: WHO/EDM/TRM/2002.1). Also available online: http://whqlibdoc.who.int/hq/2002/WHO_EDM_TRM_2002.1.pdf

Whyte, M. K. (1978). *The Status of Women in Preindustrial Societies*. Princeton, NJ: Princeton University Press.

Wikipedia (Online) Wikipedia: The Free Online Encyclopedia located at: http://en.wikipedia.org/wiki/Main_Page

Wickham, L. (Online). 'Devil's Claw for Back Pain.' *Positive Health: Complementary Health Magazine*. Online at: http://www.positive health.com (Retrieved: March 21, 2006).

Williams, M. (Online). 'Eleutherococcus Senticosus Maximum Shrub.' *Positive Health: Complementary Health Magazine*. Online at: http://www.positive health.com (Retrieved: March 21, 2006).

Winters, C. (1983). 'The Classification of Traditional African Cities.' *Journal of Urban History*, 10 (1), 3–31.

Woloch, N. (2000). *Women and the American Experience*. Boston: McGraw-Hill.

Wrigley, C. (1996) *Kinship and State: The Buganda Dynasty*. Cambridge: Cambridge University Press.

Young, C. (2001). 'Nationalism and Ethnic Conflict in Africa.' In M. Guibernau and J. Hutchinson (eds.), *Understanding Nationalism*. Cambridge: Polity.

Yusuf, A. B. (1975). 'Capital Formation and Management Among Muslim Hausa Traders of Kano, Nigeria.' *Africa: Journal of the International African Institute*, 45 (2), 167–182.

Zimbabwe (On-Line). 'Literature and Culture of Zimbabwe.' http://www.scholars.nus.edu.sg/landow/post/zimbabwe/art/greatzim/gz1.html.

Index